T0326555

Tobacco Control
and Tobacco Farming

Tobacco Control and Tobacco Farming

Separating Myth from Reality

Edited by
Wardie Leppan, Natacha Lecours
and Daniel Buckles

ANTHEM PRESS
LONDON · NEW YORK · DELHI

International Development Research Centre
Ottawa • Cairo • Montevideo • Nairobi • New Delhi

Anthem Press
An imprint of Wimbledon Publishing Company
www.anthempress.com

This edition first published in UK and USA 2014
by ANTHEM PRESS
75–76 Blackfriars Road, London SE1 8HA, UK
or PO Box 9779, London SW19 7ZG, UK
and
244 Madison Ave #116, New York, NY 10016, USA

A copublication with
International Development Research Centre
PO Box 8500
Ottawa, ON K1G 3H9
Canada
www.idrc.ca / info@idrc.ca
ISBN 978-1-55250-582-3 (IDRC ebook)

British Library Cataloguing-in-Publication Data
A catalogue record for this book is available from the British Library.

Library of Congress Cataloging-in-Publication Data
Tobacco control and tobacco farming : separating myth from reality / edited by Wardie
Leppan, Natacha Lecours, and Daniel Buckles.
p. ; cm.
Includes bibliographical references.
ISBN 978-1-78308-293-3 – ISBN 1-78308-293-3
1. Tobacco–Government policy–Case studies. 2. Crop diversification–Case studies.
3. Agricultural diversification–Case studies. I. Leppan, Wardie, editor. II. Lecours,
Natacha, editor. III. Buckles, Daniel, 1955– , editor.
[DNLM: 1. Tobacco Industry–economics. 2. Tobacco. 3. Agriculture–economics.
4. Internationality. 5. Public Policy. 6. Tobacco Use–prevention & control. HD 9130.6]
HD9130.6.T63 2014
338.1'7371–dc23
2014029204

Cover image courtesy of Sandy Campbell, IDRC.

ISBN-13: 978 1 78308 293 3 (Hbk)
ISBN-10: 1 78308 293 3 (Hbk)

This title is also available as an ebook.

CONTENTS

LIST OF FIGURES, TABLES AND PHOTOGRAPHS

Figures

Tables

Photographs

FOREWORD

Tobacco farming and particularly tobacco farmers have been and continue to be used by the tobacco industry to slow down and even completely stop progress in tobacco control. This has been achieved largely through cleverly packaged messages which have built the myth that tobacco control is targeting tobacco farming and farmers. This book separates the myths from the reality and contains important facts about tobacco control and tobacco farming, as well as the fact that viable alternatives to tobacco farming exist. The International Development Research Centre (IDRC) has been at the forefront of supporting operational research that is generating the much-needed evidence that tobacco control does not target tobacco farming or farmers. This book is a testimony to that vision and commitment.

IDRC supports research especially in low-resource countries. This research generates evidence that answers critical public health questions. These answers then lead to policy change that benefits public health and subsequently the society at large. It is a good investment. This book tackles pertinent questions and provides answers through evidence. These questions include the perception that:

- Enacting laws that reduce the demand for tobacco will negatively affect producers, particularly smallholder farmers;
- Tobacco control has a negative impact on foreign exchange earnings;
- Tobacco farming is a very lucrative cash crop particularly for smallholder farmers and viable alternatives do not exist.

In seven chapters plus an introduction and a conclusion, this book takes us through what creates tobacco leaf demand, including the manipulation of leaf prices to the benefit of the tobacco industry and detriment of the farmers. This book exposes the true negative impact of tobacco farming in low-resource countries largely as a result of cheaper labor, fewer environmental and labor restrictions and weak government oversight of tobacco farming. It then delivers us to the reality of sustainable and economically viable alternative solutions to farming. From Bangladesh to Brazil, through Kenya where the peasant tobacco farmer can now breathe better through bamboo farming, the evidence is telling. Alternatives not only exist, but their positive results to the environment, health and society can be seen relatively quickly. Small-scale farmers need support through: friendly policies that are effectively implemented; programs that support diversification from tobacco farming; and infrastructure that provides technical inputs at the farm level and access to markets for alternative crops. For alternative solutions to be successful, this issue needs to be addressed as an important part of the development agenda.

The book also reminds us that, although demand reduction measures are reducing the prevalence of tobacco users, population growth is resulting in higher absolute numbers of tobacco users. There is no rapid decline in global demand for tobacco leaf. There is still much demand reduction work to be done. It is also clear that tobacco is not a lucrative crop for small-scale farmers because it is labor intensive with serious negative environmental, health and social impacts.

As someone who has been involved in tobacco control for a while now, and who has been part of the efforts to make viable alternative livelihoods accessible to tobacco farmers at country, regional and global levels, I can say that the evidence presented in this book is worth reading, that the diversification experiences documented in this book are worth emulating and that the farmers who find themselves growing tobacco are not the problem. We also must applaud IDRC for investing in generating evidence on the effects of tobacco farming, for documenting and making the case for alternative solutions and for gathering such stellar professionals to deliver a book of this nature. I am confident that this effort will result in policy change and will ensure that many farmers are saved from the bondage of tobacco farming.

This book is worth reading!

Ahmed E. Ogwell Ouma
Regional Adviser for Tobacco Control in the WHO Regional Office for Africa

PREFACE

The genesis of this book dates back to 1994 when the International Development Research Centre (IDRC) first became involved in supporting research for tobacco control in low- and middle-income countries (LMICs). From the very beginning, IDRC saw tobacco control as not "just" a crucially important health issue but also a broader development one. IDRC recognized that the rapid growth in consumption, brought about in part by changing trade regimes, had implications for a country's economy and autonomy in setting public-health regulations. Moreover, the growth in tobacco-related diseases was putting great stress on already over-stretched health systems and negatively impacting economic productivity. At the household level, tobacco use was found to exacerbate poverty due to the financial burden of addiction, healthcare spending and loss of productive life years.

In addition, the bulk of the world's tobacco production had shifted to LMICs. Despite the claims of the tobacco industry, early evidence was beginning to show that tobacco farming appeared to be a very hard way to make a living for small-scale farmers. The evidence suggested that:

- Tobacco farming was extremely labor intensive with farming families providing much of the unpaid labor (including, in many cases, children).
- The tobacco plant leached nutrients from the soil and required large amounts of pesticides. Other environmental impacts included severe deforestation in areas where the tobacco was flue-cured or smoke-cured.
- In addition to the health hazards from the pesticides, others included smoke inhalation from tending to the drying kilns, "green tobacco sickness" from picking wet leaves, inhalation of tobacco dust from storing the dried leaves in the homestead, etc.
- Economically, while providing farmers with much-needed cash, they more often than not appeared to find themselves tied into a vicious debt bondage cycle with the tobacco companies.

As early as 2000, IDRC's then Research for International Tobacco Control (RITC) program had as one of its five research themes "Tobacco farming and alternative livelihoods." This received added impetus in 2004 during the negotiations for the WHO Framework Convention on Tobacco Control (which entered into force in 2005). Tobacco farming then came to be seen as a barrier to tobacco control as many countries were convinced by the argument that tobacco-control measures would be a disaster for their economies and small-scale tobacco farming.

This book is based on the results of research funded by IDRC, drawing from nearly 20 research projects in LMICs including: Bangladesh, Brazil, China, Ghana, India, Kenya, Lebanon, Malawi, Uganda, Vietnam, Zambia and Zimbabwe. The three largest projects (in Bangladesh, Kenya and Malawi) all initially set out to assess farmers' experiences with tobacco farming, their reasons for farming tobacco and their desire to switch to alternatives or not. Building on that, the research then shifted to working with those same farmers to assess possible alternatives, test and compare the viability of other crops and their contribution to livelihoods. While the Kenyan project aimed to replace one cash crop with another, in Bangladesh the focus was on assessing how to make the transition to diversified food production. In Malawi, the aim was to help tobacco farmers diversify to limit their reliance on tobacco but not necessarily to replace tobacco altogether in the short term. (This project also attempted to take into account other sources of farmer vulnerability such as the impact of climate variability.) Smaller projects tended to focus on more specific issues such as occupational health impacts, the knowledge, attitudes and beliefs of tobacco farmers and the impact of subsidies on tobacco farming.

In 2010, IDRC's Natacha Lecours undertook a comprehensive review of the literature and the work supported by IDRC in order to get a better sense of what was known to date and the research gaps that remained.[1] It also provided an important background document for an IDRC-sponsored international workshop, "Consolidating the Research on Alternative Livelihoods to Tobacco Farming in LMICs," in June 2011. This workshop brought together the majority of the active researchers in this field (32 people from 17 countries, largely LMICs). The objectives were to:

• Identify what was known and what research gaps still existed regarding the health, environmental and socioeconomic impacts of tobacco production;
• Identify enablers and barriers to transitioning out of tobacco production;

---

1   Part of this review was subsequently published in the journal *Tobacco Control* (see: N. Lecours, G. E. G. Almeida, J. J. Abdallah and T. Novotny. 2012. "Environmental Health Impacts of Tobacco Farming: A Review of the Literature." *Tobacco Control* 21(2): 191–196) and the comprehensive review was updated and included in this volume (see Chapter 4).

- Develop a list of research priorities to address key gaps;
- Plan a dissemination strategy of knowledge to date.

The health impacts of tobacco production in LMICs was one of the knowledge gaps identified. While some research results on the topic from high-income countries could be extrapolated to LMICs, participants felt that the impacts were likely to be even higher in LMICs given the different working conditions and the lack of protective equipment. IDRC-supported research is now ongoing in this area and initial results appear to confirm that concern.

The workshop results also indicated that more evidence was needed on the national determinants of demand for tobacco leaf. While the global demand for leaf continues to grow (something participants also felt there was a need to reconfirm), farmers faced fluctuating demand and prices for their leaves at the national level, which the tobacco industry blamed on tobacco-control measures. A number of studies were subsequently commissioned to assess these issues and form the first section of this book.

It became clear to the participants at the workshop that there was an urgent need to address the confusion and acceptance of many of the myths surrounding tobacco farming as propagated by the tobacco industry and that producing a book that consolidated what was known to date was sorely needed. This book is a response to that expressed need.

Since this book is based largely on work supported by IDRC, as well as a number of IDRC-commissioned studies aimed at addressing the knowledge gaps identified by the workshop participants, it does not cover all possible countries and issues. The biggest limitation is that big players like India and China are not studied in detail. (At the time of writing, there was an IDRC-funded project ongoing in India that was looking at the options for diversification from tobacco farming, bidi rolling and tendu leaf plucking but it was not completed in time to be included here.) However, we feel that the book covers the key arguments, based on sound scientific evidence, that help to clear the confusion surrounding this issue and remove an important barrier to the implementation of effective tobacco-control policies.

We would like to thank: all the participants at the June 2011 workshop; the IDRC research partners and the farmers with whom they worked, whose research informed this book and contributed to building the knowledge base in this field; Daniel Buckles for diving wholeheartedly into the project, contributing to the subject matter and doing much of the heavy lifting in editing; Jad Chabaan for providing guidance to the leaf-demand studies in the first section of this book; the contributors to this book who worked with us through multiple revisions; Greg Hallen, the Program Leader of IDRC's Non-Communicable Disease Prevention program, who provided much help with the Introduction and Conclusion sections and good-naturedly put up with our obsession in producing this book; Francis Thompson of the Framework Convention Alliance for much useful advice; Nola Haddadian of IDRC's Communications Division who guided us through the process of publishing; the three anonymous peer-reviewers who provided positive feedback and useful suggestions; and, of course, many thanks to the fine folks at Anthem Press for their support in publishing this book.

Wardie Leppan
Senior Program Specialist, Non-Communicable Disease Prevention
International Development Research Centre

Natacha Lecours
Program Management Officer, Non-Communicable Disease Prevention
International Development Research Centre

Introduction

SEPARATING MYTH FROM REALITY

Wardie Leppan, Natacha Lecours and Daniel Buckles

Knowledge of the harmful effects of tobacco use has prompted sustained efforts to regulate the industry that produces and markets tobacco products. The last decade has been encouraging with the development of the first global treaty negotiated under the auspices of the World Health Organization (WHO). The Framework Convention on Tobacco Control (FCTC), adopted by the World Health Assembly on 21 May 2003, entered into force on 27 February 2005 and counted 178 adhering parties as of April 2014. It is an evidence-based treaty that sets out objectives and principles that parties must follow. The articles of the convention include demand reduction measures like tax increases, health warnings, advertising bans and smoke-free environments. The FCTC also includes measures related to farming of tobacco (Article 17 on provision of support to economically viable alternative activities) and the environment (Article 18 on protection of the environment and the health of people engaged in tobacco cultivation and manufacture). As one of the most widely embraced treaties in the United Nations' history, it is a powerful tool to curb tobacco use across the globe. It is also viewed as a serious threat by the tobacco industry, which regularly challenges government implementation of the FCTC.

That the tobacco industry actively seeks to delay, dilute and defeat attempts at tobacco control should not come as a surprise. For the industry, opposition is an existential imperative. Denial and subterfuge started as far back as the 1950s when increased information on the negative health effects of tobacco use first began to appear (Proctor 2012; WHO 2008; Cunningham 1996). During the two following decades, the industry continuously challenged the veracity of the links between tobacco and disease until the scientific evidence was so overwhelming that it could no longer be questioned. It then shifted its emphasis by portraying tobacco as a major contributor to national economies and tobacco control as a threat to industry jobs, farmer livelihoods and government revenue.

This book focuses on the implications of tobacco control for tobacco farming and the myths perpetuated among policy makers and the media by the tobacco industry and front groups such as the International Tobacco Growers' Association (ITGA). Despite evidence to the contrary, industry representatives continue to say that:

• Measures to control tobacco use will suppress global demand and drive down prices for tobacco leaf thereby provoking a livelihood crisis among tobacco farmers;
• Tobacco farmers are currently relatively prosperous and that tobacco farming poses no significant risks that cannot be mitigated;[1]
• There are currently no economically sustainable alternatives to tobacco farming for small-scale farmers, particularly in low- and middle-income countries.

Considerable effort and expense have gone into asserting these claims, including sponsorship of a global campaign to engage tobacco farmers in the lobby against implementation of the FCTC (Assunta 2012). Misinformation and criticism have been directed at FCTC Article 18, dealing with protection of the environment and the health of people engaged in tobacco cultivation and manufacture, and at FCTC Article 17, which states that:

> Parties shall, in cooperation with each other and with competent international and regional intergovernmental organizations, promote, as appropriate, economically viable alternatives for tobacco workers, growers and, as the case may be, individual sellers.

While the task of promoting economically sustainable alternatives to tobacco will not be easy, the industry's description of the scope and nature of the challenge is inaccurate. As with so many previous tobacco industry attempts to block tobacco control, industry opposition to the FCTC relies

---

1   This reluctant recognition by the industry of the risks involved in tobacco farming is a relatively new development. It arose recently in a study funded by British American Tobacco (BAT) which concluded that tobacco is no worse than any other industrial crop (Pain et al. 2012). The study fails to do justice to the unique occupational health and environmental impacts of tobacco farming, or the harsh realities of tobacco farming for most farmers in low- and middle-income countries (see Lecours, this volume).

on seeding controversy and creating a fear of tobacco control among policy makers where there should be none. The industry has managed to do this by presenting Article 17 as an admission by the parties that the application of demand reduction measures of the treaty will suddenly and dramatically threaten farmers' livelihoods. This has created the false perception among some governments that parties must offer alternative livelihood programs for tobacco farmers *before* any further action is taken to control tobacco use in their country. Nothing could be further from the truth.

Evidence-based responses to claims that tobacco control is a threat in the long term to government revenues and jobs have already proven beyond a reasonable doubt that those concerns are groundless or misleading (Warner 2000; World Bank 1999; Warner and Fulton 1995). Numerous studies on tobacco taxation in both developed and developing countries show that increases in taxes on tobacco products such as cigarettes are both very effective in reducing the prevalence of tobacco use (particularly among youth) and of little or no negative consequence for government tax revenue (Kostova et al. 2011; Barkat et al. 2012; Chaloupka et al. 2000). Research shows that tobacco tax increases have actually led to increased government revenue in the short to medium term. This is due to demand for cigarettes being relatively inelastic as a result of the addictive nature of the product – for every one percent increase in price there is less than a one percent decrease in demand (Chaloupka et al. 2012).

Assessments of tobacco control show that the reduced burden of disease and premature death resulting from lower levels of tobacco use can also bring governments significant net economic benefits. Shafey et al. (2009, 43) note that, "Tobacco's estimated USD 500 billion drain on the world economy is so large that it exceeds the total expenditure on health in all low- and middle-income countries." Tobacco use imposes significant opportunity costs for already over-stretched health systems and other public services of value to all citizens. For example, the direct economic costs attributable to tobacco use in Malaysia, estimated at USD 922 million in 2008, could have funded the entire national rural development program the following year (Eriksen et al. 2012). At the household level, the high cost of addiction to tobacco products hits the poor disproportionately and directly exacerbates poverty. For example, low-income households in Egypt spend over 10 percent of household income on tobacco products (Nassar, 2003). Meanwhile, some 11.3 percent of total healthcare expenditure in Egypt is used to treat tobacco-related illnesses (Eriksen et al. 2012). In India, 25 percent of families in which a member suffers from cardiovascular disease (the leading cause of death in the country) experience catastrophic expenditures and 10 percent are driven into poverty (Mahal, Karan, and Engelgau 2010).

Research on the impact of potential job losses due to tobacco-control measures also puts tobacco industry arguments about the economics of tobacco in perspective. Germany and the Netherlands are currently among the top cigarette manufacturers and exporters. These countries could easily absorb displaced workers into other productive sectors without significant transition costs or adverse effects on overall economic activity. Making the transition to a tobacco independent economy in China presents a more difficult challenge due to the large number of consumers and producers, but even in this case tobacco manufacturing jobs in China are relatively inconsequential. They account for only 24 out of every 100,000 workers (Eriksen et al. 2012).

Potential job losses in India and Bangladesh, both countries with a tobacco industry and a large proportion of bidi smokers, are policy concerns but for reasons different from those posed by the tobacco industry. Research on bidi-dependent livelihoods in Bangladesh (Roy et al. 2012) shows the many shortcomings of bidi-manufacturing as a source of employment. The majority of bidi workers are women and children classified as unpaid assistants. They work under extremely poor conditions both in factories and at home, for financial returns that keep them relegated to "the 40% of the Bangladeshi population living below the international poverty line of USD 1.25 per day (Roy et al. 2012, 314)." Virtually all bidi users are also poor, yet spend almost 10 percent of their daily income on tobacco. This study, and research in India (Panchamukhi et al. 2008), suggests that redirecting spending from smoking to basic needs would not only have positive health benefits for smokers but would also stimulate other economic sectors and generate alternate employment for bidi workers desperately seeking a way out of an exploitative industry. Moreover, the overall economic contribution of the bidi industry at present is small and any job losses due to higher excise taxes would be temporary (Nandi et al. 2014). Research in high-income countries (Allen 1993; Buck et al. 1995) also shows clearly that reduced consumption of tobacco products happens gradually and has no lasting impact on employment or the economy as a whole because income and expenditure also move gradually to other sectors of the economy.

While the economic implications of tobacco control for government revenue and employment in manufacturing are now well understood, industry claims regarding implications for farmers have not been sufficiently challenged. This book counters the misleading claims coming from the tobacco industry by reframing the questions policy makers should be asking and providing clear and positive answers. It shows that pitting tobacco farmers against tobacco-control policies and the legitimate search for better livelihoods is actually a false dilemma promoted by the tobacco industry with a single purpose – to undermine the national and international consensus on the urgency of tobacco control as both public health and development policy.

Overview of the Book

The book is organized around the three claims made by the tobacco industry noted above, and counteracting arguments. The first section addresses the claim that "Measures to control tobacco use will suppress global demand and drive down prices for tobacco leaf, thereby provoking a livelihood crisis among tobacco farmers." As the literature shows, despite substantial progress in implementing demand reduction measures in many countries, there is no indication of an impending rapid decline in global demand for tobacco leaf. While a number of high-income countries (HICs) have been successful at reducing consumption, this has been more than offset by the growth in consumption in low- and middle-income countries (LMICs) where new trade regimes, weaker regulations, aggressive tobacco marketing and the rapid growth in their populations have provided for an expansion of the market. A recent study (Mendez et al. 2013, 50) estimates that in the absence of substantial new tobacco-control measures "the global number of smokers [...] will likely increase by 10%, to a staggering 872 million smokers in 2030, from 794 million in 2010." In this light, the first section of the book examines the actual determinants and likely evolution of demand and prices for tobacco leaf globally and in specific tobacco-growing countries, with particular attention to the structure and balance of power along the leaf marketing chain. Chapter 1 by Jad Chaaban sets the scene by examining broad trends in the tobacco leaf market and factors driving global demand. It illustrates that the above industry claim could only be true if tobacco farmers were predominantly producing tobacco not suitable for export and supplying a domestic market in which consumption was declining rapidly due to strong demand reduction measures implemented over a short period of time – a scenario not seen nor likely to be seen in any country. In reality, the overwhelming majority of farmers produce for the global market, hence demand and prices for their product are largely unaffected by demand reduction measures in their own countries. Rather, farmers' common experiences of fluctuations in demand and falling farm-gate prices are explained in large part by industry market manipulation and the weak position that farmers occupy in the leaf marketing chain.

The analytical framework used in Chapter 1, based on the established practice of mapping the marketing chain for agricultural commodities, is applied in Chapters 2 and 3 to national case studies. The analysis of determinants of demand in Lebanon by Kanj Hamade and in Malawi by Marty Otañez and Laura Graen draw attention to the relationships between the various actors in the value chain, from farmers to leaf processors, cigarette manufacturers and cigarette retailers and how these relationships shape demand, leaf prices and farmer income. While both countries are unique in some ways – Lebanon is a country with longstanding government intervention in the sector and Malawi is the most tobacco-dependent economy in the world – the cases illustrate broader themes and help to explain why tobacco farmers are the weakest and most vulnerable link in the marketing chain. Continuing growth of demand for tobacco leaf has done nothing more than perpetuate low incomes and dependency among these farmers.

Chapter 4 by Natacha Lecours takes on the notion that tobacco farming is a good way to make a living, in the second section of the book. Drawing on published sources from around the world, it systematically describes the harsh socioeconomic, health and environmental realities of growing tobacco in LMICs, and helps demystify the industry claims regarding the profitability of tobacco farming and its economic value to the national economy. The evidence shows that tobacco farmers in many contexts struggle with low net gains, high levels of indebtedness and the heavy burden of hazardous work borne by the entire family and proving particularly stressful to women and children. Occupational health hazards in tobacco farming are among the most severe in agriculture and include some problems unique to growing tobacco. Environmental impacts of tobacco farming go far beyond the immediate farm setting, where soil degradation is often severe, to the broader landscape where a host of ecosystem disruptions occur. These observations concerning the negative impacts of tobacco farming, noted as well in other book chapters, underline the fact that the worldwide tobacco pandemic is not only an important health issue but also a development one.

Recent successes and new thinking about the transition from tobacco to other crops are presented in the three chapters that form the third section of the book, countering the industry's claim that no economically sustainable alternatives exist for small-scale farmers in LMICs. A detailed study from Bangladesh by Farida Akhter, Daniel Buckles and Rafiqul Haque Tito presented in Chapter 5 delves into the evolution of tobacco farming in the country and practical experiences with the tobacco-farming transition. The chapter emphasizes the contribution farmers can and do make to the development of their own solutions and the policy implications for governments concerned about food insecurity and farmer livelihoods. Chapter 6 by Jacob Kibwage, Godfrey Netondo and Peter Magati examines a project-based approach to replacing tobacco with another crop – in this case the cultivation of bamboo for local, regional and national markets in Kenya. It highlights the role of farmer organization and technical assistance to the creation of favorable conditions for new crops. Chapter 7 by Guilherme Eidt Gonçalves de Almeida offers a sharp contrast to the community-based initiatives of the previous two chapters by examining a national policy initiative in Brazil aimed at helping tobacco farmers diversify their sources of income and thereby gradually reduce their dependency on tobacco. It points as well to the political and administrative barriers to diversification strategies and the need for a territorial approach to promoting alternatives to tobacco. Collectively, these experiences show that the issues that really matter to farmers, and that should form the core of government policies and continuing engagement with tobacco farmers, are rooted not in the fear of tobacco control but rather in the continuing challenges of smallholder agriculture.

A concluding chapter by Daniel Buckles, Natacha Lecours and Wardie Leppan summarizes and broadens arguments made throughout the book regarding the true drivers of global demand for tobacco leaf, the sources of farmer vulnerability and the policy conditions needed for the emergence of economically sustainable alternatives to tobacco farming. First, it shows that the tobacco industry business model, not tobacco-control policies, is responsible for the economic dependency and low incomes experienced by tobacco growers in LMICs. This business model makes use of the international division of labor, fewer or weaker operational and environmental restrictions in LMICs and the opportunity for vertical integration in contexts where farmers are unable to organize and negotiate for better prices. It also takes advantage of the vacuum in public investment in agricultural infrastructure and services left by broader structural adjustment reforms affecting many LMICs.

Second, the chapter delves into a discussion of the broad-based and multi-stakeholder initiatives needed to regulate the most serious abuses of public resources by the tobacco industry and create the conditions for investment in economically sustainable alternatives. The authors argue that Article 17 of the FCTC provides a constructive and modest space for the tobacco-control community and health ministries to play a supportive role to ministries of agriculture, the environment, labor, finance and rural development. These actors need to take center stage as advocates for smallholder tobacco farmers and the transition to better livelihoods. Meanwhile, the development and implementation of tobacco-control policies, the core mandate of the FCTC, can and must continue confident in the knowledge that controlling demand at the national level has no negative effect on the current generation of tobacco farmers. Reducing the burden of disease and death caused by the ongoing tobacco epidemic through tobacco control and engaging tobacco farmers in the development of economically sustainable alternatives to tobacco are compatible policy goals for national governments committed to the spirit of the FCTC and the well-being of their citizens.

Policy makers, researchers and advocates interested in the topic will find summary boxes at the beginning of each section. Each of the three boxes briefly outlines the research findings that counter one of the industry's most common myths about tobacco farming. A Policy Brief on Tobacco Control and Tobacco Farming, that synthesizes the book's arguments and policy recommendations, is also included in an annex and can be used to help promote better and more integrated tobacco control.

References

Allen, R. 1993. "The False Dilemma: The Impact of Tobacco Control Policy on Employment in Canada." Ottawa: National Campaign for Action on Tobacco.

Assunta, M. 2012. "Tobacco Industry's ITGA Fights FCTC Implementation in the Uruguay Negotiations." *Tobacco Control*. DOI:10.1136/tobaccocontrol-2011-050222. 6 pp. (accessed 7 December 2013).

Barkat, A., A. Chowdhury, N. Nargis, M. Rahman, Pk. A. Kumar, S. Bashir and F. J. Chaloupka. 2012. "The Economics of Tobacco and Tobacco Taxation in Bangladesh." Report published by International Union Against Tuberculosis and Lung Disease, Paris, 53 pp.

Buck, D., C. Godfrey and M. Raw. 1995. "Tobacco and Jobs: The Impact of Reducing Consumption on Employment in the UK." York: Society for the Study of Addiction and the Centre for Health Economics, University of York.

Chaloupka, F., A. Yurekli and G. T. Fong. 2012. "Tobacco Taxes as a Tobacco Control Strategy." *Tobacco Control* 21: 172–80.

Chaloupka, F., T. Hu and K. Warner. 2000. "The Taxation of Tobacco Products." In *Tobacco Control in Developing Countries*, edited by P. Jha and F. Chaloupka, 237–72. Oxford: Oxford University Press.

Cunningham, R. 1996. *Smoke and Mirrors: The Canadian Tobacco War*. Ottawa: IDRC.

Eriksen, M., J. Mackay and H. Ross. 2012. *The Tobacco Atlas, Fourth Ed*. New York, NY: World Lung Foundation.

Kostova, D., H. Ross, E. Blecher and S. Markowitz. 2011. "Is Youth Smoking Responsive to Cigarette Prices? Evidence from Low- and Middle-Income Countries." *Tobacco Control* 20: 419–24.

Lecours, this volume.

Mahal, A., A. Karan and M. Engelgau. 2010. "The Economic Implications of Non-Communicable Disease for India." The World Bank: Health, Nutrition and Population (HNP) Discussion Paper, 115 pp.

Mendez, D., O. Alshanqeety and K. E. Warner. 2013. "The Potential Impact of Smoking Control Policies on Future Global Smoking Trends." *Tobacco Control* 22(1): 46–51.

Nandi, A., A. Ashok, E. Guindon, F. Chaloupka and P. Jha. 2014. "Estimates of the Economic Contributions of the Bidi Manufacturing Industry in India." *Tobacco Control* Published Online First, 30 April 2014; 1–8. DOI: 10.1136/tobaccocontrol-2013-051404.

Nassar, H. 2003. "The Economics of Tobacco in Egypt: A New Analysis of Demand." HNP Discussion Paper, as part of the series, Economics of Tobacco Control, Paper No. 8.

Pain, A., I. Hancock, S. Eden-Green and B. Everett. 2012. "Research and Evidence Collection on Issues Related to Articles 17 and 18 of the Framework Convention on Tobacco Control." Report published by DD International for British American Tobacco.

Online: http://ddinternational.org.uk/viewProject.php?project=21   (accessed   7
December 2012).

Panchamukhi, R. R., T. Woolery and S. N. Nayanatara. 2008. "Economics of Bidis in
India." In *Bidi Smoking and Public Health*, edited by S. Asama and P. Gupta, 167–95.
Delhi: Government of India.

Proctor, R. N. 2012. "The History of the Discovery of the Cigarette–Lung Cancer Link:
Evidentiary Traditions, Corporate Denial, Global Toll." *Tobacco Control* 21: 87–91.

Roy, A., D. Efroymson, L. Jones, S. Ahmed, I. Arafat, R. Sarker and S. Fitzgerald. 2012.
"Gainfully Employed? An Inquiry into Bidi-dependent Livelihoods in Bangladesh."
Tobacco Control 21(3): 313–17.

Shafey, O., M. Eriksen, H. Ross and J. Mackay. 2009. *The Tobacco Atlas, Third Edition*.
Atlanta, Georgia: American Cancer Society.

Warner, K. 2000. "The Economics of Tobacco: Myths and Realities." *Tobacco Control* 9: 78–89.

Warner, K and G. Fulton. 1995. "Importance of Tobacco to a Country's Economy: An
Appraisal of the Tobacco Industry's Economic Argument." *Tobacco Control* 4: 180–83.

WHO (World Health Organization). 2008. "Tobacco Industry Interference with Tobacco
Control." Report published by World Health Organization, Geneva, 46 pp.

World Bank. 1999. "Curbing the Epidemic: Governments and the Economics of Tobacco
Control." *Tobacco Control* 8: 196–201.

Section One

THE DETERMINANTS OF TOBACCO LEAF DEMAND

Tobacco Industry Myth: Measures to control tobacco use will suppress global demand and drive down prices for tobacco leaf, thereby provoking a livelihood crisis among tobacco farmers.

Research Findings:

- Overall consumption of tobacco products will actually increase for the next several decades, driven by the growth in population and rising rates of tobacco use in low- and middle-income countries.
- Consequently, the global tobacco leaf market will remain substantial enough to sustain the current generation of tobacco farmers, most of whom sell a product that goes into global markets.
- Corporate strategies of a monopolistic industry (among other factors such as government subsidies and population growth) carry much more weight in driving demand for and production of tobacco globally and in particular national contexts.
- The real source of vulnerability of tobacco farmers to fluctuations in demand and falling farm-gate prices for tobacco leaf has to do with their weak bargaining position in the leaf marketing chain and is not due to tobacco-control measures.

Chapter 1

DETERMINANTS AND LIKELY EVOLUTION OF GLOBAL TOBACCO LEAF DEMAND

Jad Chaaban

Introduction

Tobacco industry advocates argue that tobacco-control policies are the chief culprits in reducing global tobacco leaf demand, thus negatively affecting farmers' livelihoods. The argument runs as follows: tobacco-control policies lead to a decrease in tobacco consumption and therefore to a decline in global demand for tobacco leaf. This will impoverish large numbers of farmers in poorer countries that heavily rely on tobacco leaf farming.

This chapter examines trends in the global tobacco leaf market, key features of the tobacco industry and a range of factors driving global tobacco leaf demand. It shows that tobacco-control policies play a very minor role in determining short- to medium-term global demand for tobacco leaf. Population growth, income growth, cultural norms, new technology, national economic and political dynamics, government subsidies and the corporate strategies of a monopolistic industry carry much more weight in driving demand for and production of tobacco globally and in particular national contexts. By placing global tobacco leaf demand in this broader perspective, the fallacy of the industry argument against tobacco-control policies is revealed. It also highlights the real source of vulnerability of tobacco farmers to fluctuations in demand and falling farm-gate prices for tobacco leaf – their weak position in the leaf marketing chain.

Overview of the Tobacco Leaf Market

Global consumption of tobacco products

Demand for tobacco leaf is essentially derived from demand for manufactured tobacco products, predominantly cigarettes. More than 43 trillion cigarettes have been smoked in the last ten years, with more than six trillion cigarettes being sold every year. The global market for cigarettes was estimated at USD 610 billion in the year 2010 and accounted for over 95 percent of total worldwide sales of tobacco products (Euromonitor 2011; Eriksen et al. 2012).

Health organizations and financial analysts estimate the total number of smokers today at approximately 1.3 billion worldwide (Mazars 2011; Eriksen et al. 2012). Around 20 percent of the world's population smokes cigarettes. Figure 1.1 shows the evolution of global cigarette consumption in billions of sticks since manufactured cigarettes were introduced in the late nineteenth century, forecast through to 2020.

Global consumption of cigarettes has historically been highest in high-income countries (HIC) where it increased steadily until the early 1990s. Between 1990

Figure 1.1. Estimated evolution of global cigarette consumption (in billions of sticks), to 2020

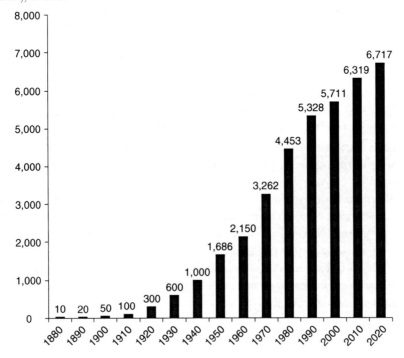

Source: Shafey et al. 2009.

and 2009, however, overall tobacco consumption diverged geographically. While in HICs tobacco demand was leveling off or slowly declining, more tobacco was used in low- and middle-income countries (LMIC). For example, during this period, cigarette consumption dropped by 26 percent in Western Europe and rose by 57 percent in the Middle East and Africa (Eriksen et al. 2012).

The shift in tobacco consumption from high-income countries to low- and middle-income countries is due to many factors. In HICs, the downward consumption trend was reinforced by changing consumer awareness of the dangers of smoking, reinforced by strong tobacco-control policies and regulations limiting advertising. In LMICs, targeted international marketing campaigns and trade liberalization driving down prices for tobacco products stimulated demand, which was sustained by higher population growth and increases in disposable income (FAO 2003a).

Figure 1.2 portrays global tobacco consumption in 2009 (Shafey et al. 2009; Mazars 2011). China alone accounted for 38 percent of tobacco consumption worldwide, followed by Russia at 7 percent. Ng et al. (2014) provide recent evidence of the impact of rapid population growth in developing countries on the prevalence of tobacco use and cigarette consumption. For instance, between 1980 and 2012, the total number of cigarettes smoked in China grew from 1 trillion to 2.3 trillion, while in the US there was a decline from 610 billion to 310 billion.

Global tobacco consumption is projected to increase steadily. The World Health Organization (WHO) estimates that over the next five years

Figure 1.2. Global tobacco consumption, 2009

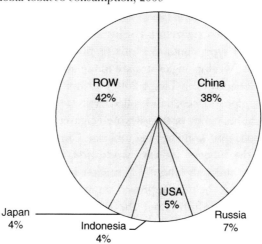

* ROW – Rest of world
Source: Shafey et al. 2009.

the number of smokers on the globe will increase at a compound annual growth rate of 3.5 percent to 4 percent. This trend will be maintained despite falling rates of tobacco use in developed countries because the population increase in developing countries will drive the overall growth in the number of tobacco users worldwide. China alone will add 8.5 million new smokers by the year 2015. Financial analysts predict that there will still be at least one billion tobacco users by the year 2050 (Euromonitor 2011).

Global production of tobacco leaf

Tobacco occupies 3.8 million hectares of agricultural land worldwide, in 124 countries. China, Indonesia, India, Brazil and Malawi account for about two-thirds of this total (FAO 2012). In 2009, some 7.1 million tons of tobacco were produced (Eriksen et al. 2012).

Tobacco will grow in any warm and moist environment. However, several factors influence the characteristics of the final product, including climate and soil conditions, harvesting methods and curing procedures. Among these, the curing method generally defines each type of tobacco (Van Liemt 2002). The most widely used curing methods are flue-curing, fire-curing, air-curing and sun-curing. After curing, which is the last stage in the production of tobacco, leaves are manufactured into the final tobacco product.

Most tobacco goes to the manufacture of cigarettes. Different manufacturers and brands use specific mixes of tobacco types (plus other additives) in their cigarettes. There are four main types of cigarettes, namely Virginia, American blend, dark and oriental cigarettes. For instance, the American blend, which is the most popular type, contains a mix of flue-cured Virginia, Burley and Oriental tobaccos. Virginia cigarettes are made almost completely from flue-cured Virginia tobacco (Van Liemt 2002). Other tobacco products include cigars, cigarillos, pipe tobacco, hand-rolled cigarettes (like bidis), roll-your-own (RYO), kretek, shisha, candy or fruit-flavored cigarettes and smokeless tobacco products like snuff, snus and chewing tobacco. Currently, almost 100 million people work in the tobacco industry worldwide, of which 40 million work in growing leaves and only about 1.2 million in manufacturing cigarettes. As discussed further below, improved manufacturing and the ongoing consolidation of the industry (not the decline in cigarette consumption) continues to drive employment down (ILO 2003).

While unmanufactured tobacco stores well, cigarettes do not. Consequently, manufacturers stock various types of unmanufactured tobacco so they can respond to increases in demand for cigarettes without delay. This practice shelters prices for cigarettes from much more variable conditions affecting the supply and prices for diverse types of unmanufactured tobacco leaf across various growing regions (Van Liemt 2002). These differences in demand and supply dynamics allow global manufacturers to seek out the lowest price for inputs and highest price for the end product.

In recent decades tobacco leaf production declined steadily in HICs as transnational tobacco corporations shifted their attention to lower cost production environments in LMICs. At the same time, support to tobacco growing in HICs in the form of subsidies and technical assistance was withdrawn and, in some cases, support was provided to farmers to switch to other crops (Cunningham 1996; Gale et al. 2000). Tobacco production shifted steadily to LMICs (FAO 2003a; Geist 2009).

Figure 1.3 depicts the world's major tobacco-producing countries. China is the world's largest tobacco producer, contributing 43 percent of global production. Three countries – China, Brazil and India – account for two-thirds of all global tobacco leaf production. The USA and the European Union, formerly major tobacco producers, currently both account for four percent each of the world's tobacco. Developing countries, including Zimbabwe, Tanzania, Malawi, Argentina, Indonesia and Pakistan, also experienced significant growth in the sector over the last decade. Percent increases in production were greatest, however, in four countries in Africa (Mozambique, Zambia, Mali and Ghana) and in Cambodia, suggesting that the diversification of sources of unmanufactured tobacco by transnational tobacco companies continues (Eriksen et al. 2012).

Figure 1.3. Tobacco-producing countries, 2011

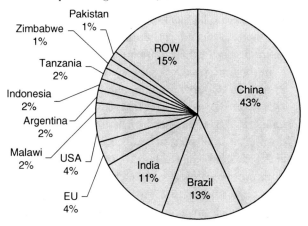

* ROW – Rest of world
Source: FAO (2012). FAOSTAT Production dataset. Online: http://faostat3.fao.org/home/index.html (accessed 1 August 2012).

Global tobacco leaf trade

International tobacco trade involves both tobacco leaves as raw material and manufactured tobacco products such as cigarettes. The total tobacco leaf trade is estimated at an annual value of USD 7 billion, whereas the annual value for international trade of manufactured tobacco products is estimated at more than twice that amount, USD 15 billion in 2010 (Mazars 2011).

Figure 1.4 shows tobacco leaf exports by country for 2011. While the European Union is still the single largest exporter of manufactured tobacco products (mainly cigarettes) and a major exporter of tobacco leaf, Brazil leads among exporters of unmanufactured tobacco leaf. China, India and the USA follow. Malawi and Zimbabwe, while smaller players in the global trade of tobacco leaf, rely very heavily on tobacco leaf exports as a proportion of the national economy. These relatively small exporters consume only a fraction of their production, and do not have an internationally competitive cigarette production industry that can compete internationally for market share (Streatfield 2005).

The largest share of tobacco produced in LMICs is ultimately traded in global markets and is therefore largely unaffected by demand reduction measures in their own country. There are a few exceptions, however. China, which produces the largest amount of tobacco leaf by far, manufactures and consumes much of its production. Nevertheless, it still accounts for 6 percent of global tobacco trade. Similarly, India is a major consumer of lower quality bidi products manufactured nationally and a major global exporter of higher quality unmanufactured tobacco and manufactured tobacco products. On balance, tobacco leaf production in both countries is likely to continue to gain from increases in overall global demand for tobacco leaf. Argentina is also an exceptional case as tobacco farmers there receive direct subsidies from taxes on local consumption. Reduced consumption nationally would eventually result in lower subsidies for farmers, depending on changes in tobacco taxes.

Illicit trade in tobacco products also stimulates the global market for unmanufactured tobacco. Cigarettes are among the most commonly smuggled products globally – 580 billion sticks were traded worldwide on the black market in 2010 (Euromonitor 2011). Financial analysts expect that the world illicit cigarette market will grow by 60 percent between 2010 and 2015, with the Middle East, Africa and Australasia being the largest growing regions. While high taxes on tobacco products are commonly cited by the tobacco industry as a major factor of smuggling, illicit trade is most prevalent in LMICs where taxes are generally low. Some LMICs have high smuggling rates and high prevalence of illicit trade, even though the price of cigarettes is low (World Bank 2003). In most high-income countries, the prevalence of smuggling is minimal even though cigarette prices are high. Ineffective sanctions against smuggling, weak border controls, organized

Figure 1.4. Tobacco-exporting countries, 2011

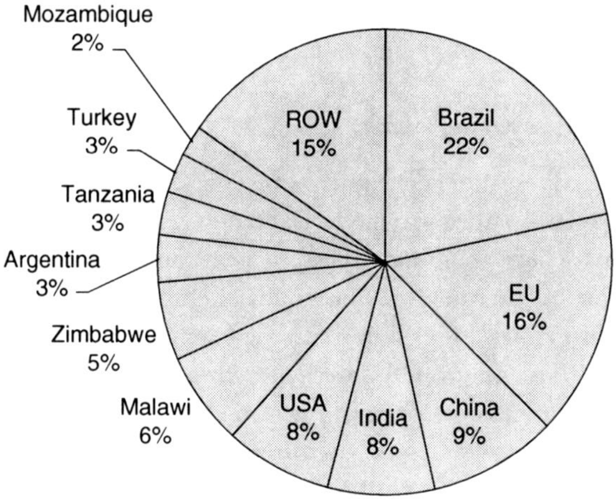

Mozambique 2%
Turkey 3%
Tanzania 3%
Argentina 3%
Zimbabwe 5%
Malawi 6%
USA 8%
India 8%
China 9%
ROW 15%
Brazil 22%
EU 16%

Source: FAO (2012). FAOSTAT Production dataset. Online: http://faostat3.fao.org/home/index.html (accessed 1 August 2012).

crime, fraud and complicity of officials with the industry seem to be the main drivers, rather than price (see Joossens and Raw (2012) for a recent discussion).

Global tobacco companies

The global tobacco industry is controlled by five multinational corporations, consolidated in recent decades through mergers and take overs. These are Philip Morris International, Altria/Philip Morris USA, Japan Tobacco International, British American Tobacco and Imperial Tobacco. As of 21 March 2012 the World Lung Foundation reported on its website that "estimates of revenues from the global tobacco industry likely approach a half trillion U.S. dollars annually. In 2010, the combined profits of the six leading tobacco companies was USD 35.1 billion, equal to the combined profits of Coca-Cola, Microsoft and McDonald's in the same year. If Big Tobacco were a country, it would have a gross domestic product (GDP) of countries like Poland, Saudi Arabia, Sweden and Venezuela."[1]

In addition to these privately-owned transnational corporations, the state-owned China National Tobacco Corporation (CNTC) is a major player. It is the world's largest tobacco company by volume and exports Chinese brands to other countries. In 2010, China produced 41 percent of manufactured cigarettes globally, followed by Russia (7 percent), the USA (6 percent), Germany (4 percent) and Indonesia (3 percent) (Eriksen et al. 2012).

---

1   World Lung Foundation: www.worldlungfoundation.org (accessed 21 March 2012).

Over 500 cigarette factories around the world manufacture about 6 trillion cigarettes yearly, with a 13 percent increase in the last decade. Most of these factories are located in China and in Europe (especially Germany).

Tobacco leaf marketing chain

Figure 1.5 describes the generic features of the tobacco leaf marketing chain. Tobacco companies procure tobacco leaves either through integrated leaf operations where manufacturers buy leaf directly from farmers or via an intermediate party, the leaf merchant. Leaf dealers or merchants are leaf-buying companies that link between farmers and tobacco product manufacturers. Two major US-based merchants dominate the global leaf-buying market: Universal Corporation and Alliance One (formerly Dimon and Standard corporation before their merger in 2005).

Tobacco leaves are internationally traded either through the auction system or the contract system. In both cases, the market is controlled by a few major companies – tobacco manufacturers and leaf processors. Under the auction system, farmers take their crop to the trading floor at the end of the growing season. Leaf-buying companies examine the quality of the leaves and establish the grade. Price is then decided by auction. In many countries, however, only one or two companies control the majority of the market so they have considerable influence over price. Evidence of price collusion among leaf buyers also suggests that bargaining power is highly concentrated (see Otañez and Graen, this volume).

Figure 1.5. Marketing chain for tobacco leaf products

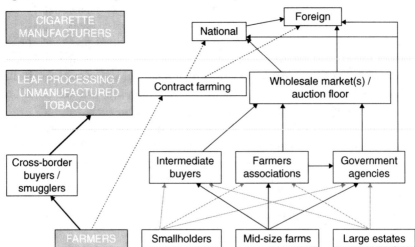

Over the last few decades, the global tobacco industry has moved in the direction of vertically integrated operations where a single company or its subsidiaries has control over all steps in the supply chain, usually through pre-established contracts with farmers. Under the direct contracting system, buyers purchase their tobacco leaf from farmers at harvest time. Buyers usually extend credit and technical support to farmers at the beginning of the year, which often ends up in a form of debt bondage allowing tobacco companies to further control farmers (see Lecours, this volume and Akhter et al., this volume). Under both systems, farmers do not have bargaining power in relationships with the small number of powerful buyers who in reality control the pricing.

Determinants of Global Tobacco Leaf Demand

As shown above, the global demand for tobacco leaf basically follows the global demand for cigarettes. Many factors influence demand for cigarettes, although key to the product's appeal is the addictive nature of nicotine, leading to habit formation and addiction. The increasing awareness over the past decades of the dangers of tobacco use has triggered international organizations and a number of national governments to change their perspective on tobacco farming and consumption and to deploy efforts to reduce tobacco prevalence. These include pricing, excise taxes and anti-smoking policies and marketing campaigns. Corporate advertising and lobbying strategies have also been curtailed. Given this attention, it is not surprising that tobacco-control policies and regulations have a strong profile in the industry and in the minds of the public. They are not, however, prominent in determining global tobacco leaf demand.

Demographic, socioeconomic and cultural factors

The prevalence of the use of tobacco products is strongly influenced by demographic trends (younger populations) in different countries and regions, access to new income and cultural factors. Overall tobacco prevalence worldwide, and consequently global demand for tobacco leaf, are still on the rise, forecast to peak in 2015. It is then expected to gradually fall by 8 percent in volume between 2015 and 2050 (Euromonitor 2011). This long-term global market decline will be affected by ups and downs, with big declines in the USA, Japan and Germany and big increases in China, India, Egypt, Indonesia, Vietnam and the Philippines. China's share of world cigarette market by volume is expected to reach 50 percent by 2050. By contrast, some major developing markets like Brazil, Turkey, South Korea and Ukraine will witness significant falls due to the adoption of stringent anti-smoking legislation (Euromonitor 2011).

Cultural factors play a role (Hosseinpoor et al. 2011). Some regions and countries have a strong tobacco culture, like the Middle East, China and India.

In the case of India, chewing tobacco (gutkha) and smoking bidis are the major forms of tobacco consumption, largely outweighing cigarette consumption.

Increases in income are positively related to demand for cigarettes (FAO 2003b; Hosseinpoor et al. 2011). Development in emerging economies has led to higher levels of tobacco use, especially among the large numbers of young people entering the workforce (Kostova et al. 2011). By contrast, a slowing economy, coupled with an increase in retail price, may have a negative effect on consumer expenditure and therefore reduce tobacco demand. However, sensitivity to changes in income differs among countries, depending on the overall level of economic development and on a range of structural and cultural factors specific to each socioeconomic group (Van Liemt 2002; Hosseinpoor et al. 2011). Cigarette consumption may decline with income decline, but alternatively people may simply opt to buy cheaper cigarettes.

Government intervention

In most tobacco-producing countries, public intervention influences tobacco production and trade to some extent, although the form and level of intervention differs considerably from country to country. Some interventions depress domestic demand while others increase it. Until the late 1980s, tobacco farmers and manufacturers in low- and middle-income countries received support and loans from international bodies like the World Bank (Novotny and Mamudu 2008). Many governments still support tobacco farmers with credit facilities, inputs and other subsidies. Governments also implement various production and trade policies such as taxes on tobacco leaf production and export, trade barriers, tariffs on imported raw tobacco, export promotion strategies, etc. In quite a few countries the government invests directly in the tobacco industry through state-owned companies, thus making profit directly from the industry over and above the taxes collected on consumption and production (World Bank 2003; Hamade, this volume; Barraclough and Morrow 2010).

Growing awareness of the harms of tobacco use and the public cost of tobacco-related disease has begun to change the nature of government intervention in the tobacco industry (Van Liemt 2002). By the 1990s, the World Bank had stopped any form of support to tobacco farming and production activities. It then became one of the major players in the tobacco-control field, working closely with the World Health Organization (WHO), the International Labor Organization (ILO) and the Food and Agriculture Organization (FAO). The WHO Framework Convention on Tobacco Control (FCTC) has since 2005 become a globally recognized international treaty that establishes the mechanism to develop and sustain tobacco-control programs at the national level through multi-lateral cooperation (WHO 2009a). Many governments have gradually adopted policies to limit the harm of the use of tobacco products to their citizens.

Tobacco-control policies create restrictions on the operating environment of the tobacco industry primarily through legislative and regulatory measures. Many governments, especially in high-income countries, have increased taxation and the price of tobacco products significantly. Price is proven to directly affect demand, the youth and the poor being particularly responsive to changes in cigarette price (FAO 2003b; Kostova et al. 2011). Other forms of legislation and effective implementation of smoke-free regulations include public smoking bans, advertising and sponsorship bans, flavor bans, health warnings and restrictions on the sale and distribution of tobacco products.

Industry analysts predict that there will be further tightening of legislation in the future. Bhutan, while a tiny country, may mark a future trend. It has actively discouraged tobacco use for decades and in 2010 introduced legislation banning the cultivation, harvesting, production and sale of tobacco and tobacco products. Smoking is not allowed in public places and limits are placed on the possession of tobacco products (the permissible amounts and penalties were amended in January 2012). Flavor bans and bans on smoking in cars are also found in some countries. Uruguay imposes, through Resolution 514, a ban on brand extension, restricting tobacco companies to marketing only one type of cigarette per brand (Euromonitor 2011). Plain packaging laws implemented in Australia as of 2012 are also proving to be effective at reducing national demand for cigarette products. New Zealand, Canada, the EU and the UK are considering similar legislation (Euromonitor 2011; Plain Packs Project 2014). Emissions control legislation and the reclassification of tobacco as a drug are also under consideration by some governments. These measures, if implemented firmly by a majority of countries worldwide (including the major consuming countries), would eventually reduce global demand for tobacco leaf. Advances in tobacco manufacturing technologies and corporate practices, discussed below, have more immediate impacts.

Advances in cigarette manufacturing technologies

Technology plays a major role in the tobacco industry. In the past century, productivity has increased significantly. As a result, employment in tobacco manufacturing has fallen. For example, cigarette machines increased capacity from 250 to 16,000 cigarettes per minute. By 1998 the UK produced

3 percent more cigarettes than it did in 1990, with 75 percent less labor (ILO 2003). New leaf processing and cigarette manufacturing techniques have also increased the filling capacity of tobacco, thereby greatly reducing the amount of raw tobacco needed for each cigarette (World Bank 2003). Wastage has also been reduced, with impacts on the demand for tobacco leaf. Additives of various kinds, including chemicals to simulate flavors, have displaced the specialized skills of leaf-blenders and made it easier for companies to produce distinctive brands with lower quality tobacco. Cigarette manufacturers substitute a cheaper tobacco type for a specific tobacco type in the cigarette blend and then technologically enhance the product to meet the characteristics of the more expensive tobacco (World Bank 2003; FAO 2003b).

Better machines, reduced waste, increased filling capacity, additives and technologically enhanced tobacco flavors are major factors driving down demand and prices for tobacco leaf, both in the past and over the longer term. These changes in cigarette production technologies help the industry lower its dependence on any one country's tobacco crop and increase their bargaining power with governments and farmers (World Bank 2003).

Corporate growth strategies

The global concentration of ownership of key parts of the tobacco leaf marketing chain has created an oligopoly of some of the most powerful corporations in the world. Despite this power, it is becoming increasingly difficult for the major industry players to thrive in high-income countries where markets for manufactured tobacco products are gradually contracting. Tobacco corporations are consequently shifting their focus to low- and middle-income countries where tobacco-control regulations and policies are less severe, production costs are lower and population is on the rise. Tobacco manufacturers consciously target these emerging markets, including youth and women, while deploying lobbying tactics in the market to delay tobacco-control legislation and regulation (Mazars 2011; WHO 2009b).

In parallel, the tobacco industry recognizes the need to diversify their product offering by catering to health conscious consumers. The quality of tobacco products and innovation in their use are increasingly becoming key elements in the corporate strategy to maintain consumers worldwide. So-called "reduced harm" product streams are constantly emerging, even though evidence suggests that these are simply a market ploy (WHO 2009b). The first to appear were the filtered and low-tar cigarettes. Both of these were subsequently proven to be as risky as regular cigarettes. This was followed by menthol-flavored cigarettes, a product with similar risks. Among non-cigarette tobacco products, moist snuff is expected to be the fastest growing (Euromonitor 2011).

Innovation in cigarette marketing today is centered on two categories of non-combustible products. One type contains tobacco (smokeless tobacco) and the other contains only nicotine (electronic nicotine delivery devices – no tobacco). Snus, ZeroStyle Mint, tobacco sticks and hard snuff "dissolvables" are examples of marketed smokeless tobacco becoming popular among users of tobacco products. Electronic nicotine delivery devices (ENDS) include e-cigarettes and nicotine replacement therapy products (gums, patches, sprays). By 2050, the market for ENDS is predicted to reach 5 percent by value of the total tobacco market and to grow faster than any other single tobacco product (Euromonitor 2011). This industry growth strategy will have a major impact on tobacco leaf demand in the future, and consequently on the livelihoods of tobacco farmers that remain in the industry.

Corporate growth strategies are driven fundamentally by cost–benefit analysis and not by concerns for the welfare of tobacco farmers. Debt bondage and the bargaining power of the big tobacco manufacturers are evidence of the disadvantaged position farmers occupy in the tobacco value chain (see Lecours, this volume and Buckles et al., this volume). Since the tobacco industry is dominated by a small number of very large companies, power imbalances are present across the market chain. Tobacco manufacturers, through the integration of phases in the marketing chain, actively manage leaf supply. Farmers in poorer countries are kept in a weak and dependent position. They have little bargaining power over tobacco leaf price on the auction floor or are often locked into fixed-terms contracts. The gradual pace of decline in global leaf demand in decades to come may provide farmers with an opportunity to break this dependency and develop more sustainable and favorable livelihoods.

Conclusions

Gradual changes in the structure of tobacco consumption worldwide and the resulting gradual decline in global demand for tobacco products will eventually affect the tobacco industry and in turn the future generation of tobacco farmers. However, it is important to note that these changes will not come about quickly and may very well be drawn out for very long periods. It is clear that overall consumption will actually increase for the next several decades, driven by the growth in population and rising rates of tobacco use in low- and middle-income countries. Consequently, the global tobacco leaf market will remain substantial enough to sustain the current generation of tobacco farmers, most of whom sell product that ultimately goes into global markets.

Concerns about the current situation of tobacco farmers are nevertheless real, albeit for reasons different from those proclaimed by the tobacco industry (see Lecours, this volume and Buckles et al., this volume). Tobacco farmers, and in particular smallholders, are the most disadvantaged link in the tobacco value chain. Corporate strategies of technological innovation and consolidation of the industry are primary drivers of fluctuating tobacco prices at the farm gate and in national and global markets. The extent of public intervention along this chain, and the lobby influence of international tobacco companies on both private and public sector actors, also affect tobacco production at a national level.

Case studies in this book on marketing chains in Lebanon and Malawi and the evolution of the tobacco sector in Bangladesh, Kenya and Brazil provide examples of how these factors work together in each context. Broadly, they can be seen as specific expressions of the evolving tobacco business model, changing production techniques and the structure and balance of power along the leaf marketing chain (see Buckles et al., this volume). These unequal relationships, not tobacco-control policies, account for major fluctuations in domestic demand and prices for tobacco leaf from one country to another.

References

Akhter et al., this volume.

Barraclough, S. and M. Morrow. 2010. "The Political Economy of Tobacco and Poverty Alleviation in Southeast Asia: Contradictions in the Role of the State." *Global Health Promotion* Supplement (1): 40–50.

Cunningham, R. 1996. *Smoke and Mirrors: The Canadian Tobacco War.* Ottawa: IDRC.

Eriksen, M., J. Mackay and H. Ross. 2012. *The Tobacco Atlas, Fourth Edition.* New York: World Lung Foundation.

Euromonitor. 2011. "Global Briefing: The Future of Tobacco." Report published by Euromonitor International, London, 48 pp.

FAO (Food and Agriculture Organization). 2003a. "Projections of Tobacco Production, Consumption and Trade to the Year 2010." Rome: Food and Agriculture Organization.

_____. 2003b. "Issues in the Global Tobacco Economy: Selected Case Studies." Rome: Food and Agriculture Organization.

_____. 2012. "The Food and Agriculture Organization Corporate Statistical Database (FAOSTAT)." Online: http://faostat3.fao.org/home/index.html (accessed 1 August 2012).

Gale, F. H., L. Foreman and T. Capehart. 2000. "Tobacco and the Economy: Farms, Jobs and Communities." Economic Research Service, US Department of Agriculture, Agricultural Economic Report No. 789. Online: http://www.ers.usda.gov/publications/aeragricultural-economic-report/aer789.aspx#.UUdEIzePxmE (accessed 17 December 2013).

Geist, H. J., K. Chang, V. Etges and J. M. Abdallah. 2009. "Tobacco Growers at the Crossroads: Towards a Comparison of Diversification and Ecosystem Impacts." *Land Use Policy* 26 (4): 1066–79.

Hamade, this volume.

Hosseinpoor, A. R., L. A. Parker, E. Tursan d'Espaignet and S. Chatterji. 2011. "Social Determinants of Smoking in Low- and Middle-Income Countries: Results from the World Health Survey." *PLOS ONE* 6(5): e20331. DOI: 10.1371/journal.pone.0020331.

ILO (International Labor Organization). 2003. "Employment Trends in the Tobacco Sector: Challenges and Prospects." Report for discussion at the Tripartite Meeting on the Future of Employment in the Tobacco Sector. Geneva: International Labor Organization.

Joossens, L. and M. Raw. 2012. "Strategic Directions and Emerging Issues in Tobacco Control: From Cigarette Smuggling to Illicit Tobacco Trade." *Tobacco Control* 21(2): 230–34.

Kostova, D., H. Ross, E. Blecher and S. Markowitz. 2011. "Is Youth Smoking Responsive to Cigarette Prices? Evidence from Low- and Middle-Income Countries." *Tobacco Control* 20(6): 419–24.

Lecours, this volume.

Mazars. 2011. "The Global Tobacco Industry." Report published by the Mazars Group. Online: http://www.mazars.ie/Home/News/Publications/Reports-Surveys/Thought-Leadership/The-Global-Tobacco-Industry (accessed 23 April 2014).

Ng, M., M. K. Freeman and T. D. Fleming. 2014. "Smoking Prevalence and Cigarette Consumption in 187 Countries, 1980–2012." *Journal of the American Medical Association* 311(2): 183–92.

Novotny, T. E. and H. M. Mamudu. 2008. "Progression of Tobacco Control Policies: Lessons from the United States and Implications for Global Action." Discussion paper published by Health, Nutrition and Population (HNP), the World Bank, Washington, DC, 68 pp.

Otañez and Graen, this volume.

Plain Packs Project. 2014. "Mythbusting." Online: http://www.plainpacksprotect.co.uk/smoking-myths.aspx (accessed 14 January 2014).

Shafey, O., M. Eriksen, H. Ross and J. Mackay. 2009. *The Tobacco Atlas, Third Edition.* Atlanta, Georgia: American Cancer Society.

Streatfield, J. 2005. "The Global Tobacco Trade: Supply and Demand Constraints in the North and South." In *Agricultural Exports as Engine of Growth for Developing Countries? A Case Study on International Tobacco Trade*, edited by World Trade Institute, 9–16. Bern: University of Bern.

Van Liemt, G. 2002. "The World Tobacco Industry: Trends and Prospects." Working paper published by the International Labour Organization, Geneva, 37 pp.

WHO. 2009a. "WHO Report on the Global Tobacco Epidemic: Implementing Smoke-Free Environments." Report published by World Health Organization, Geneva, 136 pp.

_____. 2009b. "Tobacco Industry Interference with Tobacco Control." Report published by World Health Organization, Geneva, 39 pp.

World Bank. 2003. "The Economics of Tobacco Use and Tobacco Control in the Developing World." A background paper for the high level round table on tobacco control and development policy organized by the European Commission in collaboration with the World Health Organization and the World Bank, Washington, DC, 20 pp.

Chapter 2

TOBACCO LEAF FARMING IN LEBANON: WHY MARGINALIZED FARMERS NEED A BETTER OPTION

Kanj Hamade

Introduction

The opening statement of a 2011 pamphlet produced by the state-owned tobacco monopoly Régie Libanaise des Tabacs et Tombacs (hereafter called the Régie) reads, "The tobacco crop has become a symbol of resilience, resistance and people's attachment to the Nation's land [author's translation] (2011a, 1)." This statement seeks to characterize tobacco farming as a heroic struggle against Israeli occupation, a role it did play in the border villages of southern Lebanon for more than two decades. It masks, however, the continuous manipulation of tobacco farmers by national political elites, the fundamental economic irrationality of the tobacco industry in Lebanon and the shortcomings of development policies in Lebanon's rural areas. Moreover, the positive image invoked by the statement feeds into and reinforces the lobby by international tobacco companies against tobacco-control policies in Lebanon.[1]

This chapter presents a more balanced view of tobacco farming in Lebanon by drawing attention to the historical and current political economy of the industry and the perverse logic of a trade deal between the state-owned tobacco monopoly and other actors in the supply chain that perpetuates tobacco farming. It triangulates information and data collected from published sources, from in-depth and semi-structured interviews with key informants and from the Régie's own unpublished statistics. It also maps the tobacco supply chain and examines the regional dimensions of tobacco farming. In doing so the chapter demonstrates that tobacco-control measures

---

1 Cigarette advertising is still not banned in Lebanon, despite international norms.

at the national level, even if they were to meet all the recommendations of the Framework Convention on Tobacco Control (FCTC), would have no meaningful impact on tobacco leaf production and farmer's livelihoods in Lebanon. The dynamics of tobacco production in Lebanon march to a different drummer, and cannot be reduced to economic factors only. The policy recommendations, outlined in a concluding section, call on the Ministry of Finance to lead farmers in a transition out of tobacco farming by diverting state subsidies currently provided to the tobacco industry towards implementation of a systematic rural development strategy.

The Political History of Lebanon's Tobacco Monopoly

Tobacco leaf production in Lebanon must be considered first and foremost from a political economy perspective. Beginning with the period of the French Mandate (1920–1943) and continuing to the present day, the tobacco monopoly has been one of the very few rural policy instruments created and implemented by successive Lebanese governments. Furthermore, all major changes in the tobacco industry have taken place at times of change in the political sphere.

Zamir (2000) and Firo (2003) document the use of the tobacco monopoly as a political tool during the French Mandate, from 1920 to 1943. The monopoly was inherited by the French from the Ottoman Empire, which had used it to control tobacco trade with Europe in all of its territories from the mid-nineteenth century onwards. The tobacco trade was critical to the development of the port at Beirut and to Mount Lebanon, a quasi-autonomous district within the Ottoman Empire. In 1883, the monopoly was turned over as a concession to the French-owned Régie Co-Intéressée des Tabacs de l'Empire, signaling a time of greater intervention in the region by European powers. After World War I and the creation of the Lebanese state, the monopoly remained within the French Mandate territories (Lebanon and Syria). Production grew dramatically until the end of the monopoly in 1929, a political decision that resulted in a collapse of local tobacco prices in the early 1930s.

In 1935, the French re-established the monopoly by granting exclusive rights on all Lebanese territory to a private company, the Société Anonyme de Régie Co-Intéressée Libano-Syrienne de Tabacs et Tombacs.[2] This allowed the administration of the region to grant licenses to landlords and elites in exchange for political support. Production shifted quickly from

---

2   The decision sparked important protests against the French Mandate that later led to Lebanon's independence in 1943.

Figure 2.1. Map of Lebanon showing tobacco leaf production areas

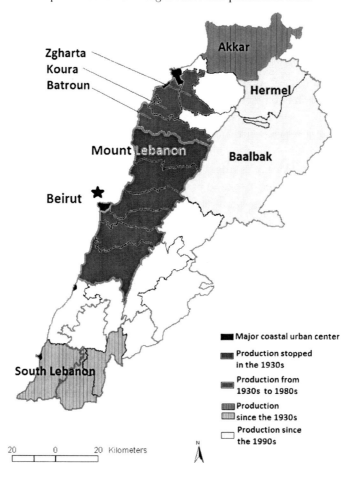

Mount Lebanon to South Lebanon and the districts of Batroun, Koura and Zghrata in the North Governorate (Figure 2.1).

Chehabist agricultural and rural development reforms

After Lebanon gained independence in 1943, new governments continued the Ottoman and French policies of focusing resources on the development of Beirut and Mount Lebanon, giving little attention to outlying rural areas. This urban and central bias deepened the inequality of economic development among regions of the country and created a situation that became more difficult to manage in the face of growing rural poverty and rural–urban migration (Traboulsi 2007). In 1958, after a major political crisis that translated into armed insurrection against the government, the head of the army, General Fouad Chehab, was elected president. He began an important series of reforms that built state institutions (a public university, Central Bank, National Social Security funds, etc.) and put in place many economic reforms.

Traboulsi (2007) argues that there was a need to rebalance a Lebanese economy dominated by an archaic banking sector and service sector. The era of "Chehabist reforms" tried to redistribute wealth initially created by the growth of the service sector and thus gain political support from the middle classes and rural populations. The reforms also aimed to mitigate the social inequality that had fueled the 1958 insurrection in the first place. In 1959 the ruling class tackled – for the first and only time in the country's history – significant policy issues pertaining to agriculture and rural development. Under these policies new state entities were established (for example, the Green Plan Directorate, a sort of rural development department within the Ministry of Agriculture) and irrigation projects funded (for example, the Litani River Dam).

Following independence, the tobacco monopoly had opened its doors to new shareholders, including the Egyptian Government and Lebanese investors. Under the Chehabist reforms, the Régie Libanaise des Tabacs et Tombacs kept its monopoly status, which allowed it to control all aspects of tobacco leaf production, the trade in manufactured tobacco products and the distribution of tobacco products. However, the reforms fixed the level of profit by the company to four percent of the sector's output. It also introduced price subsidies on tobacco leaf, taxes on tobacco consumption and custom duties for the import of tobacco leaf. These statutes were renewed every year until 1964, when the monopoly itself was renewed for another 10 years. This occurred after an agreement between the government and the Régie, in which the latter agreed to provide licenses to small-scale farmers and not just to landlords with large estates. These arrangements continued *de facto* throughout the Lebanese civil war that spanned 1975 to 1991.

Tobacco as a tool in resisting Israeli occupation

With the return of peace in 1993, the Lebanese Government assessed all of the remaining postwar administrations and special status companies, such as the Régie, oil refineries, wheat and sugar beet offices, etc. The Régie was then fully nationalized under the auspices of the Ministry of Finance and all previous licenses cancelled. New criteria gave all households permanently living in rural areas and farming on their own land or on leased land the right to obtain a tobacco license. This process of "democratization" of tobacco licenses was motivated by political aims related to the Israeli occupation of southern Lebanon and postwar reconstruction.[3] Three should be highlighted:

- Access to a tobacco license provided households in South Lebanon with sources of cash and income that would allow them to remain in their villages without having to collaborate with the occupation.
- The promotion of tobacco production in the Beqaa Governorate (Baalbek and Al-Hermel districts) provided a means to support an alternative to cannabis cultivation, which had become widespread during the civil war years.
- Making tobacco-farming licenses available in the North Governorate (especially the Akkar district) sought to support economic development within a predominantly Sunni Muslim area. This was a way to balance faster development in regions elsewhere with a predominantly Shiite Muslim population. Due to the dry climatic conditions in the region, only *tumbac* varieties of tobacco can be cultivated, which limits the commercial value of the crop.

The reforms immediately reduced the dependence of farmers in South Lebanon on landlords and local elites for access to tobacco-growing licenses. This reflected the broader national political interest in supporting a population under occupation, overriding long-standing local political elite control. The situation in Beqaa and North governorates was different, however. There, license distribution remained subject to the practice of clientelism – the exchange of licenses and services for political support. Meanwhile, in the Batroun, Koura and Zgharta districts of the North Governorate with access to better economic opportunities for farmers close to the country's core, the importance of tobacco farming decreased.

These diverging developments were the latest in the process of shifting tobacco production from site to site under the influence of both political goals and economic factors – from Mount Lebanon to South Lebanon and some districts of the North Governorate after the 1930s and later into occupied territories and areas with illegal farm production. This history also shaped the development of the tobacco leaf supply chain, a matter to which we now turn.

---

3   Israel occupied South Lebanon from 1978 to 2000, on an area equivalent to 12 percent of the country.

The Tobacco Supply Chain: Production and Trade

The fact that the Lebanese tobacco sector has evolved within a state monopoly renders the identification of its main actors – the Régie, tobacco farmers and international tobacco companies – a relatively straightforward exercise. The Régie is the main actor in the tobacco sector and represents the point of convergence in the tobacco supply chain (Figure 2.2). As a state-owned monopoly, it is in charge of all aspects of the tobacco industry in Lebanon. It is also the only economic actor directly involved in both production (through the purchase of leaf and manufactured tobacco products) and trade (through the control of local sales of tobacco products and sanctioned export/import activities).

The Régie also acts as a trade intermediary between farmers and international companies that manufacture tobacco products. It is in charge of distributing licenses for farming tobacco leaf and sets the purchase price of tobacco. It buys farmers' production subject to terms including a ceiling on the amount of tobacco leaf it will purchase from each license holder. The price paid is well over the average price paid to farmers in other national production environments, effectively subsidizing tobacco farming in Lebanon. The Régie then sorts production in its own facilities according to the quality of the product and the variety of tobacco leaf. The sorted tobacco leaf is then resold to international tobacco companies for the manufacture of tobacco products such as cigarettes. Unlike contract farming systems found in other parts of the world (see various chapters, this volume), the international tobacco companies operating in Lebanon do not influence tobacco production directly, other than through the provision of occasional technical support and production inputs (improved tobacco seed) to the Régie. Paradoxically, and for reasons discussed below, the international tobacco companies routinely pay a higher price for the tobacco they buy from the Régie than for tobacco available on the global market.

Finally, the Régie acts as a trade intermediary between international tobacco companies and distributors of manufactured tobacco products. It buys cigarettes and other tobacco products from international companies at a price slightly lower than regional (Middle East) prices. It then sells the imported goods to licensed distributors. The monopoly also sets the trade margins and profit margins for the distribution and retail of manufactured tobacco products. As a state-owned enterprise, it also interacts with the Ministry of Finance, which sets and collects both import and consumption taxes, including a Value-Added Tax (VAT) on trade. To these roles can be added the role of manufacturer of tobacco products. The Régie produces a domestic brand of cigarettes – Cedars – and tobacco used in water pipes (*tumbac*). It even imports Virginia leaf tobacco for its manufacturing operations.

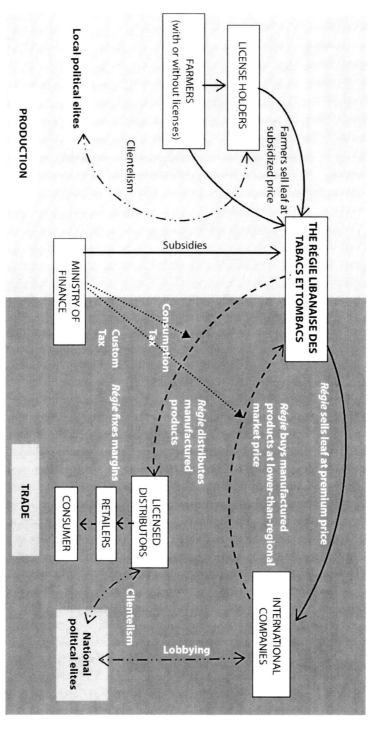

Figure 2.2. The tobacco sector's supply chain in Lebanon

The Régie sells the products in the domestic market and also exports some of its *tumbac* production. Thus, the Régie exercises monopoly control on both sides of the supply chain – production and trade.

While easily identified, the relationships between the different actors in the supply chain are extremely complex. As discussed above, historically the licensing of farmers has been subject to both national political dynamics and local clientelism through political and economic elites. What follows is a more detailed look at the current political economy of tobacco production in Lebanon, including the role of the international tobacco companies and how these relationships combine to undermine the development of a national consensus on tobacco control and block the development of better options for farmers.

Producing Tobacco: The 2006 war and its aftermath

The Régie controls the amount of tobacco leaf produced in Lebanon by defining the number of licenses issued to farmers. After the reforms and redistribution of tobacco licenses in 1993 (see above), very few new licenses were issued and lost licenses were not renewed. This in effect limited the scale of state obligations to the highly subsidized arrangements with farmers. Interviews with stakeholders and observations by the author suggest that approximately half of all licenses are actually held by non-farming license holders, who receive rent from the farmers in exchange for their right to sell to the Régie.

The Régie also controls tobacco production by setting production targets and purchase conditions. The license limits the license holder to four *dunums* (1 *dunum* = 0.1 ha) of land under tobacco production, and sets the maximum quantity of tobacco leaf it will buy at 400 kg per year for each license. In order to keep their licenses active, license holders' annual production must not fall below 200 kg for more than two consecutive years. License holders sell tobacco leaf to the Régie at fixed prices not indexed to inflation. The prices are the same throughout the country and differ only according to quality assessed by its agents at the time of collection (see Annex, Table A1).

The number of active licenses remained relatively stable between 2000 and 2005 in all three current tobacco-growing regions. A slight drop occurred in 2006 due in part to the effects of the Lebanese–Israeli war in July 2006 (see Annex, Table A2). On average, the number of active licenses decreased by 5.4 percent in the South where the conflict was greatest, 1.4 percent in Beqaa and 3.1 percent in the North. Interviews with administrators at the Régie brought out two specific reasons why licensees did not sell tobacco leaf during this period. First some farmers were not able to grow tobacco and other crops because of cluster bombs in the fields. Second many farmers planted tobacco but did not harvest it because they turned to postwar job opportunities linked to reconstruction immediately following the cessation of hostilities.

The July 2006 war and its aftermath can also be illustrated with production data (Figure 2.3). Prior to 2006, average Lebanese production was 8,433 tons per year. Production stood at an average of 7,080 tons per year after (and including) 2006, a drop of 16.0 percent. Regional effects of the war varied. In South Lebanon production dropped by 19.9 percent, in Beqaa by 9.6 percent and in the North by 10.6 percent.

Figure 2.3. Total tobacco leaf production by year and region in Lebanon, 2000–2011

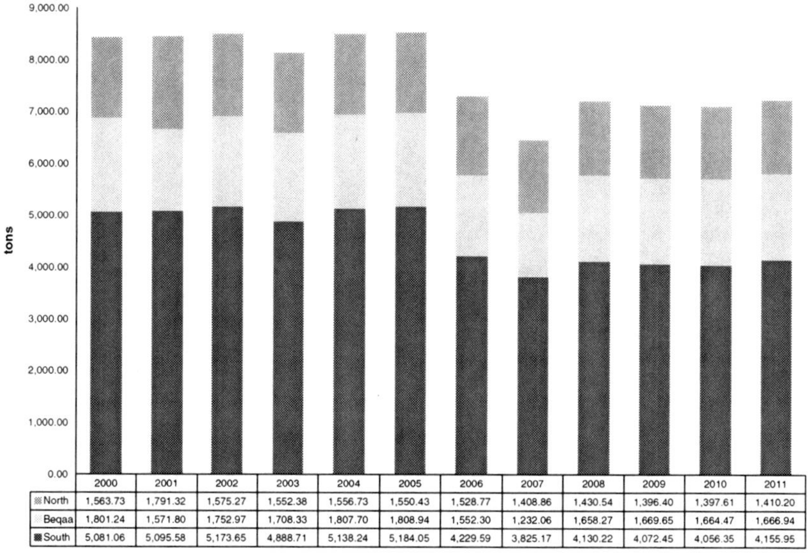

tons	2000	2001	2002	2003	2004	2005	2006	2007	2008	2009	2010	2011
North	1,563.73	1,791.32	1,575.27	1,552.38	1,556.73	1,550.43	1,528.77	1,408.86	1,430.54	1,396.40	1,397.61	1,410.20
Beqaa	1,801.24	1,571.80	1,752.97	1,708.33	1,807.70	1,808.94	1,552.30	1,232.06	1,658.27	1,669.65	1,664.47	1,666.94
South	5,081.06	5,095.58	5,173.65	4,888.71	5,138.24	5,184.05	4,229.59	3,825.17	4,130.22	4,072.45	4,056.35	4,155.95

Source: Author's adaptation from the Régie's unpublished data.

The quantity of tobacco leaf supplied to the Régie per license holder also fell at the same time but not by as much (Figure 2.4). Average production per license fell by only 15.40 percent in South Lebanon, 8.56 percent in Beqaa and 8.30 percent in the North, compared to deeper drops in total leaf production.

To explain the difference between the significant drop in production and the smaller drop in production per license holder in each region we need to consider that approximately half of all tobacco license holders are not tobacco farmers. They lease their license to the real farmers, receiving rent in exchange for their right to sell to the Régie. The 2006 drop even in areas not directly affected by the war (the North and Beqaa), and the stabilization of production at these lower levels, suggests that after 2006 some farmers in

Figure 2.4. Average quantity of tobacco sold per license holder, by region, 2000–2011

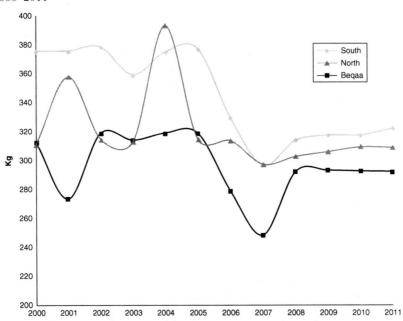

Source: Author's adaptation from the Régie's unpublished data.

all regions did not return to tobacco production. This left lease holders with less product to sell.

This observation reveals a third factor affecting farmer decision-making after the 2006 war: tobacco farming had become less attractive due to the effects of inflation. As noted earlier, the price paid to tobacco farmers and the amount of tobacco they could sell remained relatively unchanged between 1996 and 2012, even though inflation was high after 2006.[4] The inflation threat to the relationship between tobacco farmers and tobacco production has forced the Régie to reconsider its purchase policies, and in 2012 increase its longstanding purchase price.

Trading tobacco: The Régie as a trade intermediary

Tobacco farmers in Lebanon do not have direct relationships with international tobacco companies. The Régie acts as a trade intermediary, buying from

---

4   Inflation rates for this period: 2004: 1.26 percent; 2005: −0.71 percent, 2006: 5.57 percent; 2007: 4.05 percent; 2008: 10.76 percent; 2009: 1.21 percent. Source: International Monetary Fund (2001) World Economic Outlook.

farmers at one price and selling to international tobacco companies at another. What is striking about the relationship is that the purchase price is higher than the selling price, producing a net loss for the Régie from this part of their business. Consider the following trade price structure for license holders (Table 2.1) and international tobacco companies (Table 2.2) in effect for the last decade.

Table 2.1. Average buying and selling prices (USD/kg) set by the Régie

Year	South		Beqaa		North		Average	
	Buys	Sells	Buys	Sells	Buys	Sells	Buys	Sells
2000	7.94	6.27	6.28	1.79	6.39	0.78	7.40	4.14
2001	7.93	5.49	6.15	2.24	6.33	0.78	7.37	3.92
2002	7.87	5.04	6.14	1.79	6.36	0.11	7.35	3.47
2003	7.81	5.49	6.12	2.35	6.41	0.11	6.74	3.81
2004	7.57	5.60	6.06	2.35	6.10	0.90	6.92	4.03
2005	7.86	5.26	6.03	1.79	6.17	0.90	7.28	3.70
2006	7.74	3.81	6.46	2.46	6.07	0.90	7.12	3.14
2007	7.84	4.59	5.94	1.79	6.06	0.22	7.18	3.14
2008	7.85	4.59	6.38	1.90	6.02	1.01	7.15	3.36
2009	7.91	4.59	6.47	1.68	6.03	1.12	7.21	3.25
2010	8.00	N/A	6.61	N/A	6.15	N/A	7.31	N/A
2011	8.57	N/A	7.06	N/A	6.67	N/A	7.85	N/A

Source: Author's adaptation from the Régie's unpublished data.

Table 2.2. Buying price structure set by international companies (USD/kg), 2011

Type	1st Class	2nd Class	4th Class	Scraps	Class Unique (mix)	1st Class-Beqaa	2nd Class-Beqaa	4th Class-Beqaa	Class Unique Beqaa	Mix 1st & 2nd Class Beqaa
Price	6.90	5.00	3.15	0.53	5.60	3.60	2.60	1.80	2.20	2.80

Source: Régie data accessed by the author.

The Régie price structure for license holders (Table 2.1) is based mainly on quality standards (high/medium/low), with only exceptional consideration of the crop variety.[5] By contrast (Table 2.2), international tobacco companies link their price structure to the type (variety) of tobacco leaf and a different set of quality standards. As a consequence, the Régie takes a loss when it buys tobacco at a relatively uniform price across the country and sells tobacco from different sources at highly differentiated prices. The slight difference in average buying price among the three regions (Table 2.3) does not compensate for the lower selling price for lower quality production from Beqaa and lower quality and less valuable varieties from the North. This is especially true for *tumbac*, the only variety grown in the North, where export prices are as low as USD 0.11 per kg. This results in direct losses and higher levels of subsidy per kg to farmers in the North where only *tumbac* can be grown and in the Beqaa region where varieties are mixed, compared to farmers in South Lebanon who grow tobacco varieties preferred on the international market.

The Ministry of Finance provides the Régie with an annual budget for its operations, including buying, sorting, storing and then exporting tobacco production.[6] The payments cover gaps in buying and selling prices and the overall cost borne by the Régie for these operations. This includes losses due to sorting and processing of the tobacco leaf, as well as transportation and operational costs. Losses related to weight are estimated to range from 14 to 18 percent of purchased total weight, while additional costs (including transport) are estimated at USD 0.53 per kg in the South, USD 1.0 per kg in Beqaa and USD 0.50 per kg in the North.

Table 2.3 takes all of these factors into account in calculations of the amount of subsidy built into facilitating the trade relationship between farmers/license holders and the international tobacco companies. The table shows that average subsidy rates during the period 2001–2009 amounted to:

- USD 3.95 per kg, which is 50.4 percent of the buying price in the South,
- USD 5.33 per kg, which is 80.6 percent of the buying price in Beqaa (higher due to the relatively high operational costs incurred there) and
- USD 5.88 per kg, which represents 81.77 percent of the buying price in the North.

---

5   Southern Lebanon grows a variety of tobacco (Saada 6) with high demand on the international market, while Beqaa grows some Saada 6 and some local burley varieties with lesser demand. The climatic conditions in the North limit tobacco farming to the production of *tumbac*, which has less demand on the international market.

6   The Régie employs 477 full-time and 385 part-time staff, distributed in six operational centres and three procurement offices at key locations in the country. (Author's adaptation of the Régie's unpublished data.)

Table 2.3. Subsidies per region, by year (Lebanon)

Year	South			Beqaa			North			Total value
	Subsidy per kg (USD)	As % of buying price	Value (mil. of USD)	Subsidy per kg (USD)	As % of buying price	Value (mil. of USD)	Subsidy per kg (USD)	As % of buying price	Value (mil. of USD)	(mil. of USD)
2001	3.80	48%	19.36	5.30	86%	8.33	5.90	80%	10.57	38.26
2002	4.10	52%	21.21	5.60	91%	9.82	6.50	88%	10.24	41.27
2003	3.50	45%	17.11	5.20	85%	8.88	6.50	96%	10.09	36.08
2004	3.70	49%	19.01	5.20	86%	9.40	6.00	87%	9.34	37.75
2005	4.10	52%	21.25	5.70	94%	10.31	5.60	77%	8.68	40.25
2006	4.60	59%	19.46	4.70	73%	7.30	5.30	74%	8.10	34.85
2007	4.10	52%	15.68	5.50	93%	6.78	6.20	86%	8.73	31.19
2008	3.80	48%	15.69	5.30	83%	8.79	5.30	74%	7.58	32.07
2009	3.90	49%	15.88	5.50	85%	9.18	5.30	74%	7.40	32.47

Source: Author's estimates from the Régie's unpublished data.

The average nominal value of subsidies over the same period was USD 36.02 million per year (50.80 percent of the total went to farmers in the South, 24.30 percent to those in Beqaa and 24.90 percent to farmers in the North). Compared to the period before 2006, the average nominal value of subsidies decreased by 15.6 percent after 2006.

The largest part of the cost of these subsidies is borne by the national treasury, through the Ministry of Finance. International tobacco companies share the cost to some extent, however, by paying more for Lebanese tobacco than they would on the international market. Tobacco accessed from Turkey, Macedonia or Bulgaria, where volumes are much greater and transaction costs lower, would be much cheaper. For example, in 2000, the Régie sold Lebanese-produced tobacco leaf to international companies at an average price of USD 4.14 per kg when world prices for a similar product were estimated to be USD 3.00 per kg (Jaffee 2003). As discussed further below, the amount of tobacco leaf purchased by companies is relatively small and serves mainly as a means to justify access to the large Lebanese market and illegal regional market for manufactured tobacco products.[7]

The Régie has a similar way of working. It too suffers a financial loss from trade in tobacco leaf in Lebanon but achieves a positive net revenue once profits from trade in manufactured tobacco products are accounted for. Chaaban et al. (2010) estimate that the 2008 net revenue of the Régie was USD 50.3 million.

Understanding the Trade Deal Between the Régie and International Companies

Philip Morris International, Altadis USA, British American Tobacco and R.J. Reynolds Tobacco Company are active in the Lebanese market, buying small quantities of tobacco leaf and selling large amounts of manufactured cigarettes. They buy an oriental type of tobacco leaf from the Régie and mix it with American tobacco varieties to produce cigarettes, which they sell on the national and international market.

Although part of Lebanese production is considered to be of high quality,[8] the international companies' demand for Lebanese tobacco leaf stems from the fact that entry into the Lebanese market for their manufactured

---

7   Administrators in the Régie reported during interviews with the author that international companies have even purchased Lebanese tobacco leaf without requesting its shipment, thereby treating it as a complete loss.

8   Top Quality Saada 6 (approximately four to five percent of tobacco leaf production) is sold at USD 6.90 per kg (2011) to international companies (see Table 2.4).

tobacco is – informally – conditioned by the purchase of Lebanon's domestic production. The Régie, as the trade monopoly, buys manufactured tobacco from international companies at a price kept secret at the request of the companies.[9] In addition, the Régie buys Virginia varieties of tobacco leaf from the companies to create the blend the Régie uses to produce a domestic cigarette brand (Cedars). Table 2.4 outlines the quantities and value of the tobacco trade in Lebanon (with the exception of the low-quality *tumbac*, which is exported by the Régie without involvement of the international tobacco companies).

Table 2.4. Lebanese tobacco production, 2008–2012

Year	Manufactured product (value in 1000s of USD)		Tobacco leaf (value in 1000s of USD)		Trade balance in 1000s of USD		Total
	Import	Export	Import	Export	Manufactured product	Tobacco leaf	
2008	158,727	111	5,021	33,726*	−158,616	28,705	−129,911
2009	176,814	659	1,070	18,576	−176,155	17,506	−158,649
2010	223,323	12	4,985	22,059	−223,311	17,074	−206,237
2011	265,663	54	694	20,867	−265,609	20,173	−245,436
2012	328,397	84	10,542	20,353	−328,313	9,811	−318,502

* In 2008, export of tobacco leaf was higher than production because of the delay in export the Régie experienced as a result of the July 2006 war (2006 and 2007).
Source: Lebanese Customs website database (accessed 12 September 2012).

These transactions are only part of the picture, however. In a recent paper, Nakkash and Lee (2012) discuss the "complicity" of transnational tobacco companies in smuggling cigarettes into Lebanon – a situation that has occurred due to weak governance and political instability in Lebanon since the 1970s. According to the authors, smuggling aims to give international tobacco companies additional revenue from the Lebanese market by bypassing the monopoly's relatively low prices paid for the manufactured products of international companies and the high taxes on cigarette consumers. Furthermore, they argue that international tobacco companies use Lebanon as an entry point into the regional market, especially into Syria and Jordan. Cigarettes imported legally into Lebanon by the Régie are then smuggled into Syria. These dynamics probably explain a large part of an increase of

---

9   Interviews with administrators suggest that the monopoly held by the Régie has been used to negotiate a price slightly lower than the regional market price.

64 percent in the import of manufactured tobacco into Lebanon between 2008 and 2012 (see Annex, Table A3).

Illegal transactions in tobacco products seem to be operating on a large scale due to the instability in Syria. An increase in the amount of manufactured product imported to Lebanon can be read as an increase in the quantity of cigarettes smuggled into Syria – with or without international companies' complicity. This trend is likely to continue as long as the armed conflict in Syria persists, since the USA[10] and European Union's (EU)[11] trade sanctions have rendered trade with Syria relatively difficult for international tobacco companies. The "re-export" of cigarettes to Syria has to pass through Lebanon's 450 licensed tobacco distributors, which in some cases are franchises of international brands (Chaaban et al. 2010). In fact, the increase in import demand is driven – and initiated – by licensed distributors' requests for more manufactured product from the Régie.

Licensed distributers are the final link in the supply chain, before the product reaches retailers and the consuming public. The Régie also controls the sale of distribution licenses and sells manufactured tobacco only to licensed distributors. All prices and price margins are fixed by the monopoly, taking into account consumption taxes and tariffs imposed by the Ministry of Finance. When consumers in Lebanon see an increase in cigarette prices, this is the result of either an increase in taxes or a change in the trade agreements between international tobacco companies and the Régie. According to Chaaban et al. (2010), licensed distributors operate within a margin of five percent and a profit of only 1.7 percent on the retail prices. In 2010, this translated into an average profit for retailers of USD 0.07 per package. Within this framework, including a tax rate of 44 percent (all taxes included), sales of tobacco products in 2008 amounted to USD 473.3 million (Chaaban et al. 2010).

International Companies and Tobacco Policy Control

While the tobacco monopoly exercised by the Régie places international tobacco companies in a relatively weak bargaining position, evidence suggests that the companies do have an important influence on politicians when it comes to tobacco-control policy. According to the American University of Beirut's Tobacco Control Research Group (2012, 1), the Lebanese government

---

10 Since it passed the Syria Accountability and Lebanese Sovereignty Restoration Act of 2003, the US has banned all export of non-food and non-medicinal American products (defined as more than 10 percent of components made in the US) to Syria – a ban that includes manufactured tobacco.

11 EU Council's Implementing Decision 2012/256/CFSP of 14 May 2012 banned trade with the Syrian General Organization of Tobacco.

has failed to introduce tobacco-control measures because of "pressure from tobacco industry lobbyists and allies, who ensure mass tobacco advertising and promotion and prevent any successful implementation of a comprehensive tobacco advertising law in the country as well as other policies such as health warnings and public smoking bans."[12]

As demonstrated earlier in this chapter, the economic relationship between the Régie and the farmers/licensees, on one side, and between the Régie and international companies, on the other side, operates according to a non-economic logic. These dynamics confirm that tobacco-control measures in Lebanon would have no impact on the livelihood of Lebanese tobacco farmers as demand is driven by other factors and actors. Historically, political elites at the national level used tobacco-growing licenses to control the rural labor force and gain the support of local landlords (Firo 2003). They continue to use licenses today for both electoral and political reasons. Subsidies to tobacco farmers are a relatively easy way for the government to engage with farmers without having to invest in designing and implementing serious rural development policies that could help farmers improve their livelihoods and contribute to a stable Lebanese society.

For its part, the tobacco industry will likely continue to pay more for tobacco leaf so long as it can sell its manufactured products in Lebanon (and regionally). Protecting its tobacco consumer base is all that matters. The industry also benefits from the current system by using the dependency of farmers, which goes back to 1959, as an argument against tobacco-control policies. This comes at the expense of tobacco farmers who expose themselves and their families to the health risks of tobacco farming (see Lecours, this volume). By contrast, the arrangements are a win-win situation for the Lebanese political elite and cigarette manufacturers.

Tobacco Leaf Mode(s) of Production and Alternatives

Developing better options for farmers in Lebanon should be a priority for policy makers. Finding alternatives for farmers is two-pronged – it involves both a search for alternative crops and a serious attempt at developing rural parts of the country. Such a policy could contribute to lifting farmers out of

---

12 After many efforts, pressures and counter pressures from the tobacco industry and tobacco-control activists, the Lebanese Parliament passed a law banning smoking in public spaces (17 August 2012). The law was effective as of 3 September 2012. Although people are still skeptical about its application, public authorities affirm that the law will be applied strictly. For more details, refer to the Daily Star newspaper article of 18 August 2012 online: http://www.dailystar.com.lb/News/Local-News/2011/Aug-18/Parliament-passes-no-smoking-law.ashx#axzz25DpNg79y (accessed 1 September 2012).

the impoverishment they face due to the dependency relationship created by the tobacco monopoly, local political elites and the government's neglect of rural development. This section explores these issues, first by delving into the economics of tobacco production in the three tobacco-producing areas and then by exploring through a case study what alternatives might look like.

Three regions, three modes of production

In 2010, 11,094 farmers grew tobacco in Lebanon on an overall area of 8,328 ha, representing 3.43 percent of all agricultural land in the country. While this is a seemingly small percentage, Lebanon is among only five countries in the world that farm more than 1 percent of their agricultural land with tobacco (Chaaban et al. 2010). Among tobacco farmers, averaged across regions, 20.3 percent grow nothing but tobacco (Table 2.5). For some 65.7 percent of tobacco farmers, tobacco is their main crop. Furthermore, most tobacco farmers are full-time farmers (72.2 percent) and have few other ways to earn income. These features make them highly dependent on income generated by tobacco production and subsidized by the government.

Table 2.5. Tobacco farmers and land under tobacco production, 2010

Category	Only crop	1st crop	2nd crop	3rd+ crop	Total
Number of farmers	2,247	5,039	2,431	1,377	11,094
South 7,532 farmers	19%	52%	21%	8%	100%
Beqaa 1,842 farmers	21%	44%	20%	16%	100%
North 1,720 farmers	25%	21%	28%	26%	100%
Total area under tobacco (ha)	1,739.15	4,100.54	1,619.31	868.75	8,327.75
South	36%	48%	41%	30%	42%
Beqaa	38%	42%	34%	34%	39%
North	26%	10%	24%	36%	19%
Average size (ha)*	0.77	0.81	0.67	0.63	0.75
South	0.43	0.50	0.43	0.41	0.47
Beqaa	1.72	2.10	1.53	1.02	1.77
North	1.08	1.18	0.79	0.69	0.91

* While the maximum area under tobacco per license is 0.4 ha, individual farmers typically access several licenses. See below.
Source: Author's adaptation of data from FAO and the Ministry of Agriculture census 2010.

In 2010, only 48.3 percent of license holders actually farmed the land themselves that year. The data shows that license holders that also farm tobacco are more common in South Lebanon than in other regions, representing 59 percent of license holders compared to 32.4 percent in Beqaa and 38.1 percent in the North.

Differences in the average quantity of output per licensed farmer and the management of licenses also reflect regional differences in agricultural modes of production (Table 2.6). In South Lebanon, yields of tobacco per hectare are much higher than elsewhere in Lebanon (1.15 tons per ha compared to 0.51 in Beqaa and 0.88 in the North). As a result, farmers in the South can more easily meet the minimum production quota required by the Régie buyers.[13] This is reflected in the average number of licenses owned or rented by an individual farmer in the South (1.35). By contrast, farmers in Beqaa and the North must own or rent on average 2.26 licenses and 2.03 licenses (respectively) in order to meet the minimum production quota required by the Régie. When these figures are adjusted by the average quantity of tobacco sold per license the contrast is even greater. In other words, in a production environment with lower yields and lower quality of product, farmers in Beqaa and the North must manage many more tobacco licenses (through ownership and/or rental) so that they can operate within the buying regime of the Régie.

Table 2.6. Average quantity output per farmer and licenses, by region, 2010

Category	South	Beqaa	North
Average quantity of output per farmer (kg)	538.55	903.62	812.56
Required number of licenses*	1.35	2.26	2.03
Adjusted number of required licenses**	1.70	3.09	2.63

*Based on 400 kg of tobacco per license.
**Based on average quantity sold per number of active licenses per region.
Source: Author's adaptation of unpublished Régie data.

It is important to note that lower quality and yield in Beqaa are not only due to the tobacco variety and climatic conditions but also to a lack of tobacco-growing expertise. The crop was introduced to the region only in 1993 and little knowledge of the crop has been accumulated. Farmers rely on seasonal field workers whose know-how is significantly lower than that of experienced

---

13 As mentioned earlier, the license limits the license holder to 0.4 ha of land under tobacco production and sets the maximum quantity of tobacco leaf it will buy at 400 kg per year for each license. In order for license holders to keep their licenses active, their annual production must not fall below 200 kg for more than two consecutive years.

tobacco farmers in the South and they do not necessarily grow tobacco every year. These differences in the modes of production translate into different output values per license holder, per farmer and per hectare, as shown in Table 2.7.

Table 2.7. Average value of output in USD, by region, 2010

Category	South	Beqaa	North
Total output value	32,440,000	11,000,000	9,330,000
Per license holder	2,541	1,935	2,064
Per farmer	4,300	5,834	5,604
Per ha	9,274	3,388	5,894

Source: Author's adaptation of unpublished Régie data.

In summary,

- In the South, tobacco is grown by small-scale farmers that rely mainly on skilled household labor and adequate post-harvest handling to produce better quality tobacco leaf.
- In Beqaa, tobacco is grown with higher levels of inputs (drip irrigation), including larger areas of land than elsewhere (owned or leased) and unskilled hired agricultural workers.[14] These farmers focus on producing a high quantity of relatively poor grade tobacco in hopes of creating a secure income. Farmers in Beqaa also use their licenses as collateral for small loans – including loans from mainstream financial institutions – so they can finance the inputs needed for this more capital and land-intensive tobacco production system.
- In the North, tobacco production is primarily a strategy for income diversification and risk reduction, using relatively high levels of labor and extensive amounts of land. Farmers here are attracted to tobacco because the value of the crop is not subject to volatile market fluctuations.

Farmers in the three regions also share certain characteristics. Overall, tobacco farming is significantly more labor intensive than other forms of agricultural production, as shown in Table 2.8. The exception to this is in the North where farming of any kind has not been capitalized to the extent that it has in Beqaa or in the South.

---

14 Low-wage agricultural workers – mostly from Syria – are available in Beqaa and in the North. This was not the case in the South, during or after the Israeli occupation.

Table 2.8. Full-time farmers and labor intensity comparisons

Region	Tobacco farmers (%)	Labor intensity per ha*	Standard deviation	Non-tobacco farmers (%)	Labor intensity per ha	Standard deviation
National	72.20	0.4964[a]	0.32702	48.60	0.3641[a]	0.30057
South	69.70	0.5468[b]	0.32313	45.70	0.4032[b]	0.29962
Beqaa	70.30	0.3787[c]	0.34120	56.30	0.3084[c]	0.32620
North	85.10	0.3219	0.27092	47.10	0.3112	0.30416

*Labor intensity per ha = [number of full-time household workers + number of full-time hired workers + number of part-time household work + number of days of hired seasonal workers in a year] / total agricultural area.
**An independent sample t-test was conducted to compare labor intensity between tobacco farmers and non-tobacco farmers:
　(a) At the National level there was a significant difference (df= 10782; p=0.000).
　(b) In the South there was a significant difference (df= 9552; p=0.000).
　(c) In Beqaa there was a significant difference (df= 1872; p=0.00).
In the North there was no significant difference.
Source: Author's adaption of census data from the FAO and Ministry of Agriculture 2010.

Data presented in Table 2.8 also confirms that the proportion of full-time farmers is much higher among tobacco farmers than it is among other farming groups, who tend to be more diversified. There is little justification for this situation, especially in Beqaa and the North where tobacco leaf is one of many available agricultural crops that could allow people to be full-time farmers. Tobacco farming became an option in these regions not because of its inherent profitability or suitability to the region but rather because it is highly subsidized and relatively risk free. This has the effect of pushing farmers in Beqaa and the North towards tobacco cultivation, rather than dedicating their time and natural resources to food and agriculture.

In the South, the situation is somewhat different. While there is less poverty compared to the North and Beqaa (Laithy et al. 2008), agricultural villages in the South lack the same level of access to natural resources, especially water. Within the region, tobacco farmers have been marginalized from successive levels of economic development benefiting other populations. Development has included access to public administration jobs, access to higher education, migration to Beirut and, most importantly, emigration (student emigration to Eastern Europe and the Soviet Union in the early 1970s and late 1980s, middle-class emigration to the United States and Canada and traders' emigration to Africa). In other words, tobacco farmers – who were the backbone of economic policies against occupation and who helped keep occupied villages inhabited – have been and are still excluded from economic development processes in their own region. This situation is illustrated below through a village case study.

Aytaroun: Traditional farming and the unprofitability of tobacco

Aytaroun is representative of how tobacco farming occurs in the South of Lebanon. It is a frontier village where tobacco has been grown since the French Mandate period. Production significantly increased after Independence. During the Israeli occupation (1978–2000), villagers who did not flee subsisted on cash they received from remittances – from family working in Beirut or abroad – and on income from tobacco farming.

Only 9.7 percent of all agricultural land in the village is irrigated and all tobacco plantations are rain fed. In 2010, 74 percent of farmers in Aytaroun were growing tobacco leaf (51 percent on leased land). All tobacco farmers are license holders. They usually lease an additional license in order to be able to sell product quantities of more than 400 kg. According to Bazzi (2008), only 1 percent of tobacco farmers in the South farm without licenses of their own.

Tobacco farmers have land that averages 11.4 *dunums* (1.4 ha), on which tobacco is the main crop for 86 percent of farmers. The second crop is olive trees (44.5 percent) followed by hard wheat (36 percent), both of which are produced for home consumption and harvested by other residents of the villages, with whom farmers share the yield (author's adaptation of census data from FAO and the Ministry of Agriculture, 2010). In other words, the

Photograph 2.1. Children and mother sorting harvested tobacco leaf, Aytaroun, South Lebanon, 2012

Photo credit: Wael Al-Ladiki.

Photograph 2.2. Air curing of tobacco leaves in a home, Ayrtaroun, South Lebanon, 2012

Photo credit: Wael Al-Ladiki.

agricultural mode of production in Aytaroun is mostly traditional and has not yet been fully monetized.

Tobacco production starts in January with the cultivation of seedlings, an operation performed in the house backyard or on one part of a field. The farmer prepares the soil and transplants seedlings in late March and early April in order to benefit from the late April rains. Afterwards crops are not irrigated, a stress that increases the nicotine and tar content of the harvested plant. Harvest occurs during summer over a period of about two months, from mid-June to mid-August, and usually requires 50 days of labor. Stringing, hanging and curing processes happen at home with household labor (Photograph 2.1). Once the leaves are cured, they are packed and stored at home for collection by the Régie in October (Photograph 2.2).

Tobacco production is very labor intensive and usually involves all or most household members. Table 2.9 estimates levels of profit for different types of farmers in Aytaroun, based on their household labor dependency ratio[15] and on the legal status of the land they cultivate (owned or leased). This estimation

15 The household labor dependency ratio (number of persons working on the farm divided by the total number of household members) equals 1 in 43 percent of the cases – that

Table 2.9. Estimation of farmers' profit (USD) from the 2010 tobacco crop, Aytaroun, Lebanon

Category	Household labor dependency = 1		Household labor dependency not equal to 1	
			Mean=0.5	Mean=0.57
	Own land (19%)	Lease land (24%)	Own land (30%)	Lease land (27%)
Average household size	3.39	4.48	6.05	5.41
Average tobacco land size (*dunum*, 1du=0.1ha)	5.19	5.52	5.93	5.20
Average tobacco output (taken from South average per ha)	596.51	634.34	681.95	598.00
Author's estimation of cost (total value)	6331.81	8137.61	6150.09	6125.73
Leasing land cost (USD 33 per *dunum*)		182.03		171.60
Leasing license cost (25% of value of per kg above 400 kg)	393.01	468.68	563.90	396.00
Production cost (estimated at USD 150 per du, inclusive of plowing and inputs)	778.05	827.40	889.50	780.00
Total non-labor cost	1171.06	1478.11	1453.40	1347.60
Permanent labor cost (25% of time from January to June at minimum wage rate)	499.50	499.50	499.50	499.50
Household labor cost (50 days per household labor at minimum wage rate 2010)	4661.25	6160.00	4197.19	4278.63
Total estimated labor cost	5160.75	6659.50	4696.69	4778.13
Average cost according to Bazzi (2008) USD 485 per du	2515.70	2675.26	2876.05	2522.00
Income from tobacco leaf (calculated based on average price received in the South)	4772.04	5074.72	5455.60	4784.00
Profit: according to Bazzi (2008) cost calculation	2256.35	2399.46	2579.55	2262.00
Profit: author's calculation considering labor	−1559.77	−3062.89	−694.49	−1341.73
Profit: author's calculation without labor	3600.98	3596.61	4002.20	3436.40

is also compared with Bazzi's (2008) estimation.[16] The results show that the return on tobacco farming is low even if the cost of household labor is not taken into account. When the cost of household labor is taken into account, tobacco farming is not profitable.[17]

Furthermore, as shown in Figure 2.5, when Bazzi's (2008) estimation is used, the annual (average) return for tobacco farmers is equivalent to only 60 percent of the country's annual minimum wage.[18] When using the author's estimation, it amounts to 92 percent. In either case, the data suggests that the returns from tobacco farming are insufficient to fight poverty. Whether one uses the upper or lower (subsistence) poverty line, profits remain lower than the minimum required for subsistence.

Rural activists in Aytaroun report that the main reasons why local households grow tobacco leaf are:

• Because water is very scarce, it is difficult to grow other crops. Although water harvesting is widely used in Aytaroun, the amount of water collected does not allow for its use in agriculture.
• There are few other job opportunities.
• During the summer season, tobacco production employs mainly women and children, thereby providing additional income for the household.[19]
• Tobacco production is a tradition in the region, and farmers have expertise with the crop. Most have been farming tobacco for at least four generations.

---

is, all family members are involved in tobacco leaf production. The ratio differs from one for larger-sized households.

16 The author's estimations are based on semi-structured interviews with farmers in Aytaroun, and on adapted data from the FAO and Ministry of Agriculture census (2010), as well as from the Régie. Bazzi's (2008) estimation was based on a questionnaire conducted with a sample of Aytaroun farmers.

17 Labor was considered to be equal to: preparation and cultivation: six months' labor, two hours per day, for a total of 45 person-days of work + 50 labor days per number of household members working on the farm. For a household of four, with all working on the farm, total labor days per year amounts to 245 days. Accounting for the fact that the average tobacco farm in Aytaroun is 0.55 ha, total working days for a household of four is equivalent to 445 working days per ha. An unpublished World Bank study (2010) (cited by Chaaban et al. 2010) estimated total tobacco working days per hectare in Lebanon at 605 days. The author considers this figure to be an over-estimation.

18 In 2010, the official Lebanese minimum wage was USD 333 per month. The average wage in the national labor market was around USD 800 (Muhanna 2011).

19 Table 2.8 shows that most tobacco farmers are full-time farmers. However, this data must be viewed with caution. Farmers (average age is 52) do not usually have another fixed job, but often work as low-skilled day workers, something not taken into account when statistics are gathered. Furthermore, households benefit from additional income generated by the secondary breadwinners in the family.

Figure 2.5. Tobacco farming and poverty in Aytaroun, 2010

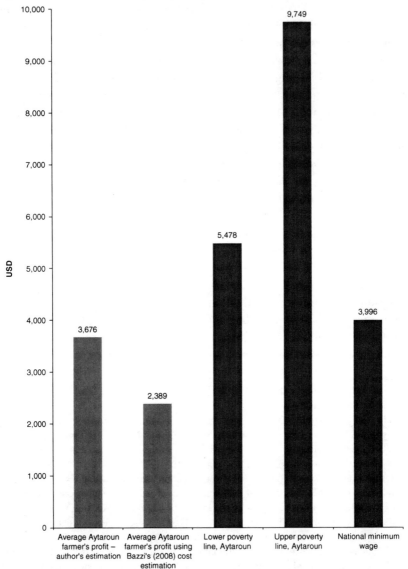

Sources: Author's adapted calculation of the poverty line based on the poverty line for the South estimated by Laithy et al. (2008), adjusted for inflation and for tobacco farmers' household size.

In terms of alternatives, olive trees could be an option, but only for farmers that actually own land. Even those who own land are not ready to make the kind of investment required to grow olives – orchards take an average of five years before they become productive. In such cases, policy interventions would be needed to subsidize farmers wanting to convert from tobacco growing to permanent olive production.

Local activists often present production of *zaatar* (a local kind of oregano) for green leaves, spices and aromatic oil as an alternative to tobacco. An Aytaroun *zaatar* cooperative exists, and has been independently sustainable since 2000. This kind of production could carry a value-added label like local, cooperative or organic. Despite this, many farmers remain skeptical due to their lack of expertise with *zaatar*, concerns about the long learning curve this kind of production required and uncertainty about markets. In the absence of policy and extension support, few tobacco farmers in Aytaroun are willing to make a step toward conversion to aromatic plant production.

Conclusion and Recommendations

Demand for tobacco leaf production – especially after it moved from the center of Lebanon (Mount Lebanon) to the peripheries (the South, Beqaa and the North) – is above all determined by political factors. The Régie's monopoly was and still is used as a political tool by the ruling classes to control the labor force in rural areas and to ensure political support from local landlords (pre-1993) or to strengthen clientelistic networks (post-1993). The political economy of the sector also reflects a lack of sociopolitical agency on the part of the farming population and limited economic opportunities in rural areas.

Analysis of the supply chain and relationships among the various actors shows that production and trade are two separate but interdependent processes. The Régie's monopolistic relationship to farmers and to international leaf buyers allows the agency to run business losses in the sphere of tobacco leaf production and recover profits in the sphere of sale of manufactured tobacco products. This creates a net gain that is to the benefit of both the national political elites and the international tobacco companies.

The political economy of tobacco subsidies and special trade deals has become the main obstacle to agricultural development in all three tobacco-growing regions. It has created a static situation in recent years in which tobacco production levels and numbers of farmers have remained almost unchanged. Production technology and forms of work organization in Southern Lebanon have remained virtually unchanged since the introduction of the tobacco leaf in the mid-1930s. The ongoing use of land-extensive technologies to produce tobacco leaf in the North and Beqaa when capital and natural resources are available also points to a farmer malaise and rural development inertia. Farmers are content with low-cost, low-risk and highly subsidized tobacco production systems. However, the analysis also shows that farmers do respond to external economic shocks, such as the high inflation rate after the July War of 2006, which prompted some farmers to diversify or abandon farming.

Tobacco farming is unprofitable when labor costs are factored in, and cannot be sustained at the level of small-scale production without being subsidized. The absence of rural development policies and the lack of political will to develop rural areas only deepens the crisis. Any policy intervention in this static sector must be well planned. A direct cut of subsidies would lead to a collapse of the system, leaving farming households, especially in the South, without immediate alternatives. Furthermore, the sudden liberalization of the market could lead to a transformation in the mode of production from small-scale family farming to large-scale capitalized cultivation of tobacco. This would be to the benefit of international tobacco companies, as it would allow them to establish direct contractual relationships with the farmers.

Despite these risks, alternatives to tobacco production are necessary if the country is to move away from the current state of inertia in rural development. The Régie, thanks to its regional infrastructure and longstanding relations with farmers, could play a positive role in supporting a transition out of tobacco. The Ministry of Finance has a self-interest in promoting this role, since buying tobacco leaf and subsidizing tobacco farming creates a financial burden for the government as a whole.

The development of an alternative is inconceivable outside the framework of a rural development strategic plan. Such a strategy should be holistic, sustainable and based on high value-added crops or labels (such as organic, local, etc.). In order to move in that direction, political will and state financing are needed. The strategy also needs to acknowledge and strengthen the human capital and resources represented by farmers and rural areas. They have a critical role to play in building a future that rejects the license exclusivity that exists in the current situation while also avoiding the re-creation of new dependency relations.

The present research comes on the heels of Lebanon's first effective tobacco-control policy, a smoking ban in all public spaces implemented in September 2012. Will this political concern for public health extend to rural development? Will it open the door to fiscal reforms in which additional taxes collected on tobacco could contribute to financing an effective and sustainable rural development strategy?

Annex

Table A1. The Régie price list based on quality of tobacco

Period	Quality and price (in USD per kg)		
	Good	Average	Low
1996–2009	10.63	7.70	3.00
2010–now	11.80	8.40	3.23

Source: Régie Libanaise des Tabacs et Tombacs (1996, 2011b).

Table A2. Number of active licenses by governorate

Region	2000	2001	2002	2003	2004	2005	2006	2007	2008	2009	2010	2011
South	13,524	13,557	13,666	13,609	13,700	13,738	12,825	12,884	13,160	12,823	12,768	12,894
Beqaa	5,759	5,744	5,497	5,438	5,665	5,669	5,558	4,957	5,673	5,693	5,686	5,699
North	5,025	5,007	5,003	4,961	3,959	4,925	4,871	4,736	4,725	4,563	4,518	4,565
Total	24,308	24,308	24,166	24,008	23,324	24,332	23,254	22,577	23,558	23,079	22,972	23,158

Source: Author's adaptation of data from the Régie's unpublished records.

Table A3. Lebanese tobacco imports and exports, 2008–2012

Year	Manufactured product (Weight in tons)		Tobacco Leaf (Weight in tons)	
	Import	Export	Import	Export
2008	10,107	17	831	10,676*
2009	11,368	185	198	6,236
2010	12,557	1	720	6,398
2011	13,779	8	693	5,276
2012	16,557	13	1,675	7,195

* In 2008, the export of tobacco leaf was higher than production because of delays caused by the July 2006 war.
Source: Lebanese Customs website database (accessed 12 September 2010).

References

American University of Beirut Tobacco Control Research Group. 2012. "The Transnational Tobacco Industry Effectively Hampers Tobacco Control Policy-Making in Lebanon. Policy Series: Tobacco Control Research 1." Fact sheet. American University of Beirut Tobacco Control Research Group, Lebanon, 2 pp. Online: http://www.aub.edu.lb/ifi/public_policy/rapp/Documents/policy_series/tobacco_control_research/20120315ifi_tcrg_policy_series_01_nakkash.pdf (accessed 4 December 2012.

Bazzi, A. I. 2008. "The Impact of Raw Tobacco Subsidies on the Rural Economy and Environment in Lebanon: Opportunities and Challenges to Income Diversification in the Case of Bintjbeil." Master Thesis, American University of Beirut.

Chaaban, J., N. Naamani and N. Salti. 2010. "The Economics of Tobacco in Lebanon: Estimation of the Social Cost of Tobacco Consumption." Research paper published by the American University of Beirut, Tobacco Control Research Group. Online: http://www.aub.edu.lb/ifi/public_policy/rapp/rapp_research/Pages/economics_of_tobacco_lebanon.aspx (accessed 4 December 2012).

Firo, K. M. 2003. *Inventing Lebanon: Nationalism and the State under the Mandate*. London: Tauris.

IMF (International Monetary Fund). 2001. *World Economic Outlook Data Base*. Washington DC: IMF. Online http://www.imf.org/external/pubs/ft/weo/2001/01/data/ (accessed 31 July 2014).

Jaffee, S. 2003. "Malawi's Tobacco Sector. Standing on One Strong Leg Is Better than on None." World Bank, Africa Region Working Paper Series No. 55. Online: http://www.worldbank.org/afr/wps/wp55.htm (accessed July 2013).

Lebanese Customs Database. 2012. Online: http://www.customs.gov.lb/customs/trade_statistics/5year/search.asp (accessed 12 September 2012).

Lecours, this volume.

Laithy, H., K. Abu-Ismail and K. Hamdan. 2008. "Poverty, Growth and Income Distribution in Lebanon." Country Study published by the International Poverty Centre, United Nations Development Programme. 24 pp.

Muhanna, I. 2011. "White Paper on the Economic and Social Implications of the Requested Minimum Wage Increase." Muhanna actuary and consultant. Online: http://www.orientation94.org/uploaded/MakalatPdf/dirasset/Salary_Adjustment_in_Lebanon_%5B1%5D.pdf (accessed 11 July 2012).

Nakkash, R. and K. Lee. 2012. "Smuggling as the 'Key to a Combined Market': British American Tobacco in Lebanon." *Tobacco Control* 17: 324–31.

Régie Libanaise des Tabacs et Tombacs. 1996. Administrative price circular. Beirut: Lebanon.

_____. 2011a. Communications pamphlet. Beirut: Lebanon.

_____. 2011b. Administrative price circular. Beirut: Lebanon.

Traboulsi, F. 2007. *A History of Modern Lebanon*. London: Pluto Press.

Zamir, Z. 2000. *Lebanon's Quest: The Road to Statehood 1926–1939*. London: Tauris.

Chapter 3

"GENTLEMEN, WHY NOT SUPPRESS THE PRICES?": GLOBAL LEAF DEMAND AND RURAL LIVELIHOODS IN MALAWI

Marty Otañez and Laura Graen

Introduction

As the world's top burley tobacco leaf producer, Malawi is at the center of global discussions on the human costs of tobacco growing and the economic implications of tobacco control. On the one hand, health activists advocate in favor of alternative livelihoods and diversified crops while publicizing the harmful effects of tobacco growing. On the other hand, tobacco industry representatives and their supporters argue that tobacco control will quickly destroy Malawi's leaf sector and national economy. Some industry advocates have even described efforts to curtail smoking as a "new form of imperialism" by health advocates against tobacco farmers, jobs and national revenue (Assunta 2012). This kind of rhetoric and more direct influence on government officials are routinely used to derail health policy making in Malawi and internationally. Not surprisingly, Malawi currently has very weak tobacco-control policies and regulations (Otañez et al. 2009; WHO 2011). Meanwhile, the perspective of tobacco farmers and farm workers on fairness and control of their own destinies is buried and out of sight.

This chapter examines Malawi's tobacco leaf sector and counters industry arguments that tobacco-control measures harm jobs, revenues and livelihoods in Malawi. We focus on the following questions: What are the structural determinants of demand for tobacco leaf in Malawi? What are the main drivers of changes in the country's tobacco leaf market (quantities and prices for tobacco leaf)? What is the relative importance of demand-side tobacco-control interventions, as compared to other factors, in shaping the evolution of tobacco leaf demand in Malawi? How do leaf-buying companies influence Malawi's market structure, through private and public sector linkages?

Malawi's tobacco supply chain has received considerable attention from researchers interested in the economic practices and processes associated with the production and export of tobacco (Prowse 2011; Tchale and Keyser 2010; Koester et al. 2004). The chapter builds on these studies by using the methods of market chain analysis (Kaplinsky 2000) and framework provided by Chaaban (this volume) to create a portrait of the production structure and the distribution of net income among the key players in the tobacco marketing chain in Malawi – tobacco tenant farmers, farmer associations, government regulators, transnational leaf-buying companies and transnational manufacturers.

The main leaf-buying companies in Malawi are Limbe Leaf (a subsidiary of US-based Universal Corporation) and Alliance One (a subsidiary of US-based Alliance One International). Philip Morris International, British American Tobacco (BAT), Japan Tobacco and Imperial Tobacco purchase tobacco leaf produced in Malawi for use in tobacco products manufactured elsewhere. The analysis of the relationship among these actors draws on information about tobacco prices, farming costs, transportation and auction charges, export prices for burley leaf which accounted for 93.8 percent of Malawi's tobacco leaf production in 2010 (Chirwa and Dorward 2011) and leaf-buyers' revenues collected from economic reports, industry assessments and government documents. These sources were supplemented and cross-checked with findings from interviews with tobacco farmers, farm workers, government authorities and tobacco industry officials.

By combining economic and ethnographic data the study is able to illustrate the tobacco supply chain from the point of view of people at the farm level, and bring to light the exploitative practices of landowners and the monopolistic practices of leaf buyers and cigarette makers. This approach differs from purely quantitative studies of commodity chains and offers another tool for integrating qualitative research into health policy studies (Otañez et al. 2006; Mathie and Carnozzi 2005).

The findings converge around a critical view of the role of tobacco companies in creating poverty at the farm level in Malawi. They show that leaf buyers and cigarette manufacturers make extra profits (economic rents) through the exercise of bargaining power and other tactics at points along the marketing chain where they have few or no costs to recover. Transactions under conditions of absolute monopoly power allow companies to extract profit above levels that would normally apply in competitive markets. The analysis also shows that the extra revenue flow is perpetuated by an industry-controlled alliance that seeks to protect itself by obstructing national tobacco-control policies and blocking the development of alternative farmer livelihoods.

The Role of Tobacco in Malawi

Malawi is the top global producer of burley tobacco leaf, accounting for an average of 18 percent of global exports of burley between 1998 and 2011 (Table 3.1). Burley leaf cultivated by farmers in Malawi is air-dried, low-nicotine and neutrally flavored. It is a "filler" type of tobacco, which companies use for Marlboro, Camel and other high-end brand cigarettes (Lea and Hanmer 2009; Jaffee 2003). The other burley type produced globally is known as a "flavor" type of tobacco because it has a stronger flavor and is similar in weight to flue-cured tobacco (Alliance One International 2006). Unless stated otherwise, this chapter refers to the filler-type burley produced in Malawi.

Until recently, farmers in Malawi sold burley leaf at auctions where leaf companies purchased the product in an open market system. In 2012, Malawi introduced the Integrated Production System (IPS) or contract farming alongside the auction system. This chapter examines both, although further study is needed to probe the impacts of the IPS as it develops in Malawi. Nsiku (2007) warns that contracts negotiated between parties of unequal power will provide leaf companies with even greater control over Malawi's tobacco marketing chain than the current auction system.

Exports from four countries (Malawi, Brazil, US and Mozambique) amounted to 59 percent of the global exports of burley leaf in 2011. Within Africa, Malawi, Mozambique, Uganda and Zambia accounted for 39 percent of global exports in 2011. Between 2007 and 2009, partly due to an expansion of leaf growing in Mozambique spearheaded by Universal Corporation, a doubling of burley production in Africa occurred (Brown and Snell 2011). Two US-based leaf-buying companies, Universal Corporation and Alliance One International, through subsidiary companies in Malawi, purchase 60 percent of Malawi's leaf. Tobacco leaf buyers sell the product through pre-arranged contracts to Philip Morris International, BAT and other global cigarette manufacturers (Otañez et al. 2007). Virtually all of Malawi's tobacco leaf is exported. While the Nyasa Manufacturing Company, established in Malawi in 2009, makes cigarettes with 100 percent locally grown tobacco, the proportion of total production used by Nyasa is nominal.

Tobacco accounted for 52 percent of Malawi's export earnings in 2010, making the country by far the most reliant on tobacco of any country in the world (Figure 3.1). In the 1990s and 2000s, export earnings from tobacco in Malawi averaged about 65 percent of total exports (Tchale and Keyser 2010).

Table 3.1. Top burley tobacco leaf exporters, 1998–2011 (million kg)

Country	1998	1999	2000	2001	2002	2003	2004	2005	2006	2007	2008	2009	2010	2011
Malawi (NP)	114	114	142	115	125	103	152	120	123	87	169	209	193	208
Brazil (KF)	84	102	95	89	116	115	144	137	133	105	100	122	90	111
US (NP)	270	251	143	156	136	124	128	93	98	102	95	91	81	71
Mozambique (NP)	2	3	6	12	19	23	39	44	46	28	45	58	55	66
EU* (PA)											65	69	55	42
Thailand (P)	40	40	28	38	42	43	45	42	40	43	36	38	38	36
China (PA)	60	60	62	55	50	45	45	29	25	25	36	35	35	37
Argentina (NP)	29	41	39	36	50	38	56	60	52	37	42	48	35	42
India (P)	9	11	9	10	6	11	9	13	12	15	21	25	21	12
Bangladesh (P)	1	0.5	0.9	1	2	2	3	4	4	5	7	14	25	16
Philippines (PA)	13	19	27	20	25	25	9	8	8	8	9	13	17	19
Uganda (P)	3	6	6	7	15	17	19	2	4	6	10	16	9	14
Guatemala (PA)	17	13	12	12	10	11	13	12	11	10	11	13	16	15
Zambia (PA)	3	3	3	4	8	12	24	22	17	5	15	15	15	14
Mexico (P)	33	26	23	20	15	17	13	11	13	10	10	10	7	14
Others	204	204	202	195	193	192	187	175	139	135	64	59	58	58
World Total	882	894	798	770	812	778	886	772	725	621	735	835	750	775

*EU countries' data reported by country or in other category prior to 2008 by Universal Corporation.
NP: Non-parties to the Framework Convention on Tobacco Control (FCTC).
PA: Partner of the Working Group on Articles 17 and 18 of the FCTC.
P: Parties to the FCTC.
KF: Key facilitator of the Working Group on Articles 17 and 18 of the FCTC.
Sources: Universal Leaf Tobacco Corporation (2007, 2008, 2012).

Figure 3.1. Exports of all tobacco types and other cash crops in Malawi, 1990–2010

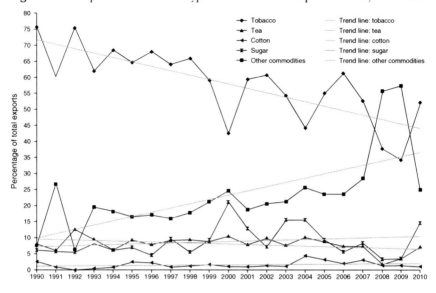

Source: Data from the Reserve Bank of Malawi (2008, 2012). Authors' estimates for 1999, due to incomplete data.

Tobacco export percentages fell from 67 percent in the first decade of this period to 50 percent in the second decade. Exports of sugar and other commodities increased between 1990 and 2010, representing a higher rate of exports among commodities that are alternatives to tobacco. Tobacco also accounts for 13 percent of Malawi's gross domestic product, contributes 23 percent of its tax revenue, and comprises 53 percent of all agricultural exports (Table 3.2). In 2011, the country earned USD 236 million from sales of 208 million kilograms of burley tobacco at an average price per kilogram of USD 1.13 (Maulidi 2011). Except for Zimbabwe (at 18.9 percent of export earnings) (Organisation for Economic Co-operation and Development 2012), virtually all other tobacco-growing countries in the world rely on tobacco for less than five percent of their export earnings (World Bank 1999).

Malawi's production of burley leaf in terms of auction sales volume increased from 95 million kilograms to 200 million kilograms between 1998 and 2011 (Zant 2012). The increase in production in the 2000s was due to the liberalization of the tobacco sector and changes in tobacco-growing policy. Bakili Muluzi, the president who took over from Kamuzu Banda in 1994, headed an administration that revised the 1964 and 1972 *Special Crops Acts* which had stated that large estates were to be the exclusive producers of export crops such as burley tobacco. These acts also stated that smallholder

Table 3.2. Economic indicators of the importance of tobacco in Malawi, 2010

Indicator	Tobacco's contribution
Foreign earnings	52%
Volume of agricultural exports	53%
Agricultural GDP	43%
Overall GDP	13%
Tax revenue	23%
Employment	780,000
Area cultivated	195,000 hectares
Proportion of area cultivated	2.95% (out of 4.6 million hectares under cultivation)

Sources: Adapted from Tchale and Keyser (2010), Otañez et al. (2007), Eriksen et al. (2012), FAO (2012) and Reserve Bank of Malawi (2012).

tobacco farmers were only allowed to grow oriental and Malawi Western, two unpopular and less lucrative tobacco types (Mkwara 2010). Under the acts, smallholder farmers that typically cultivated crops on less than one hectare were restricted to maize and other subsistence crops (Tobin and Knausenberger 1998). The Muluzi government's reforms made it possible for smallholder farmers to cultivate burley tobacco for export, which generated dramatic growth in this sector. The government initiated this agricultural reform following pressure from donors such as the World Bank and the US Agency for International Development (USAID) to liberalize the economy and alleviate poverty through liberalization of the tobacco-growing sector (Zant 2012).

In Malawi, 2.95 percent of arable land is devoted to tobacco, making it the country with the highest proportion of arable land devoted to tobacco (Eriksen et al. 2012). From 2000 to 2007, land devoted to tobacco growing increased from 194,000 hectares to 253,000 hectares, with burley accounting for 91 percent (229,000 hectares) of the crop (Malawi Government 2010).

The authors estimate that in 2010 approximately 780,000 people cultivated tobacco in Malawi. The estimate is based on data showing the total area dedicated to tobacco in 2010 (195,000 ha) and an FAO calculation (2003) that in Malawi an average of four people are needed to cultivate one hectare of tobacco. It is difficult to know with certainty what proportion of tobacco farmers fall into the major sub-categories of tobacco farmers identified by Kadzandira et al. (2004): smallholders (growing tobacco as individuals or in clubs), estate and tenant farmers (growing tobacco on large estates) and contract farmers (contracted to grow tobacco and accessing land through rental arrangements). Some studies of Malawi's tobacco-growing sector cite figures for smallholder farmers without accounting for tenant farmers and farm workers (Jaffee 2003), provide estimates of tenant farmers without stating the number of smallholder farmers (Torres 2000) or provide estimates of the number of tobacco farmers without a breakdown by farmer category (Matabwa 2012). Analysis by the authors of data from 2004 to 2012 suggests that smallholders probably represent about 60 percent of tobacco farmers, estate farmers and tenants about 30 percent and contract farmers about 10 percent (Table 3.3). As discussed below, the category of contract farmers is growing and is of concern because farmers who enter into contracts with industry agents quickly become indebted.

Table 3.3. Burley tobacco farmers in Malawi, 2010

Farmer Type	Farmers
Smallholder	468,000 (60%)
Estate / tenant	234,000 (30%)
Contract	78,000 (10%)
Total	780,000

Source: Authors' estimate from FAO (2012), Koester et al. (2004), World Bank (2005), Limbe Leaf (2006–2012) and Reserve Bank of Malawi (2012).

Despite Malawi's deep connections to tobacco growing, the country has relatively low rates of tobacco use. Smoking prevalence among adults is 10.7 percent, with adult males (20.3 percent) smoking at higher rates than females (1.5 percent) (WHO 2011). Compared to the general adult population, smoking rates among university students in Malawi are higher for both males (29 percent) and females (17.6 percent) (Kasapila and Mkandawire 2010). Tobacco growing has not led to high smoking rates in Malawi, probably due to limited household income and the absence of low-cost cigarette manufacturing in the country. It has, however, had other impacts on human health, as discussed below.

Tobacco and food insecurity

Despite decades of exporting tobacco around the world, Malawi remains at the bottom of most human development indicators (UNDP 2011 and 2012):

• 26 percent of Malawians suffer from illiteracy;
• 56 percent of the population lacks access to clean water;
• 28 percent of the population lacks access to decent sanitation facilities;

- 27 percent suffer from under-nourishment; and
- 53 percent of children under five suffer from stunting due to poor nutrition.

The causes of food insecurity in Malawi are complex and due to a variety of structural problems, including an overwhelming dependence on maize as a food source (Harrigan 2008). Diversifying household food intake from an overreliance on maize alone is critical to the food security equation. Harrigan suggests that a sensible approach to a viable food security strategy for Malawi needs to include imports, domestic production, food crop diversification, subsistence production and livelihood diversification. Nevertheless, for decades Malawi has been unsuccessful in its diversification efforts. And tobacco has failed to contribute to the solution.

Tobacco growing in Malawi is associated with stunting in children, with up to two-thirds of children on tobacco farms considered stunted, a number higher than for farms growing any other cash crop (Masanjala 2006; Wood 2011). Scope for achieving food security among smallholder tobacco farmers in Malawi is limited to households with access to more than 0.8 hectares of land. Tobacco growing requires a four-year rotation, and tenant farmers who typically farm on smaller size plots find it financially difficult to achieve food security (Orr 2000). Local farmers with sufficient land can afford to rotate crops and allow portions of land to recover by remaining fallow. For smallholder farmers, however, this is not possible. In Malawi, 48 percent of smallholders own less than 0.5 hectares of land. The labor costs needed to grow burley tobacco also erode earnings that might be used to achieve food security (Orr 2000). The high labor demands of burley tobacco, met through hiring workers who receive piece-rate earnings, also conflict directly with the labor required for maize grown in the same season.

Structural food insecurity and chronic cycles of seasonal hunger due to poor climate conditions continuously threaten tobacco and non-tobacco families and food security in Malawi (Atwell 2013; Mandala 2005). Predictors of food insecurity among rural Malawians include the total cultivated land per capita, educational attainment of heads of households, rainfall variability and household size (Fischer 2013). While systematic comparisons of food security between tobacco and non-tobacco farms are lacking, Torres (2000) found that of 1,110 tobacco tenant households surveyed in Malawi only two percent were considered food insecure, that is, eating one meal or less a day. The author states, however, that the finding conceals how this is achieved. Households that lack sufficient cash to purchase their own food obtain food advances from farm authorities. Farm authorities tend to inflate food costs and deduct debts accrued through food and other advances (for clothes and medicine) from earnings available from tobacco sales. Wait times for payment

Photograph 3.1. A pregnant woman carrying tobacco leaves from the field to her home in Kasungu, Malawi. Women and children provide much of the labor used in smallholder tobacco farming.

Photo credit: Laura Graen.

on tobacco sold can be up to three months, adding further economic stress to households. This ultimately leads to debt incurred to repay food advances and other input expenses such as fertilizer and seed (US Department of State 2011a). These conditions and related debt bondage contribute as well to the highest rate of child labor in the tobacco-growing sector found anywhere in the world (Otañez et al. 2006). Circumstances on farms, as well as tenants' lack of power to influence leaf prices, also make children and adults vulnerable to forced labor conditions and to human traffickers who take vulnerable people from the country's southern region to the central region and from Malawi to tobacco farms in Mozambique and Zambia (US Department of State 2011b).

As discussed below, poverty and the absence of meaningful development among smallholder tobacco growers at the farm level in Malawi are part of a marketing chain that operates for the benefit of a small local and national elite, leaf buyers and cigarette manufacturers. This situation justifies efforts to identify and fund alternatives to tobacco growing, an option that international development and public health agencies active in the country are beginning to explore.

The Tobacco Marketing Chain in Malawi

The tobacco marketing chain in Malawi is comprised of production and processing activities (Figure 3.2). The chain begins at the level of production with the acquisition of farm inputs. Smallholders and other farmers who cultivate tobacco use fertilizers and seeds from multinational companies such as Monsanto, and from the Agricultural Research and Extension Trust (ARET), a local organization funded through auction levies and the Tobacco Association of Malawi (TAMA). In the contract marketing system discussed later in this chapter, farmers are provided with inputs through an advance on their harvest, thereby taking over the production process from start to finish.

Tobacco production occurs in a nine-month season (September to June) and may vary each year due to climate issues such as early, late or inadequate rains. Selling at auction occurs from March to September. Farmers and laborers perform activities such as preparing nurseries for seedlings, transplanting seedlings into fields, weeding, harvesting, grading and baling. Representatives and extension agents from ARET, farmers' associations and (in the contract system) leaf-buying companies visit farms to check on yields, qualities, pesticide applications and grading practices.

Figure 3.2. Tobacco marketing chain in Malawi

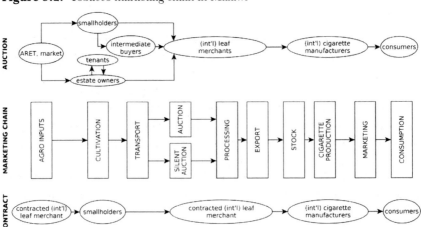

Both tenant farmers and smallholders participate in the tobacco marketing chain. A tenant farmer refers to an adult, typically a male with a spouse and children under the age of 16. Literacy rates among tenants are relatively low, contributing to low earnings and lack of knowledge of the occupational health and safety risks of tobacco farming (Lea and Hanmer 2009). Labor arrangements, typically verbal, between tenants and farm owners require that tenants grow tobacco on land provided by the owner, with the owner agreeing to pay tenants after the tobacco is sold at auction. Farm authorities such as managers and absentee landlords lend food, cash and any other basic goods to tenants, deducting these costs from tobacco earnings. Often, farm authorities inflate prices for inputs, a situation that contributes to conditions that entrap families in debt and force parents to send their children to fields instead of classrooms. Tenant and smallholder farming households account for most of the estimated 78,000 child laborers as young as age five that cultivate leaf for global markets in Malawi (Otañez et al. 2006).

Smallholder farmers are independent farmers and landholders who sell their tobacco through auction or the contract marketing system. In contrast to tenant farmers who are given land and inputs by their landlord and are forced by him/ her to grow tobacco and sell it to only him/her, smallholder farmers own the

Photograph 3.2. Tobacco tenant farmer Isaac Ching'oma lives with his wife and three children on a tobacco estate in Kasungu, Malawi

Photo credit: Laura Graen.

Photograph 3.3. Bundles of leaves being sewn before hanging to dry in a shed in Kasungu, Malawi

Photo credit: Laura Graen.

plot of land where they grow tobacco and are free to decide which crops they want to grow and to whom they want to sell. Two studies present a detailed economic portrait of Malawi's smallholder tobacco farmers. Takane (2007), in a study of 186 tobacco households, provides a breakdown of the cost structure of tobacco produced by smallholders in 2005 and estimate of net income (Table 3.4). In the sample, a farmer grows on average 0.35 hectares of tobacco, producing 749 kilograms per hectare of finished product. Costs include inputs such as seeds, fertilizer, pesticides and materials to create holding barns and sacks used for baling the tobacco leaf. Two of the largest expenses are hired labor excluding family labor (31 percent) and fertilizer (41 percent). Average net income per season for a tobacco farmer estimated by Takane (2007) was MWK 14,315 or USD 93. In a 2005 study on Malawi's tobacco sector, Jaffee and Nucifora estimated the net return to family land, labor and capital at 16 percent.

After the harvesting and drying of leaves, tobacco bales weighing about 100 kilograms each are prepared for delivery to the market. Malawi's Tobacco Control Commission (TCC) is mandated to register and provide licenses to smallholder tobacco farmers' clubs, tobacco estates and tobacco-buying companies. Also, estate and contract farmers are required to be registered with the TCC. In 2011, 10,660 farmers were registered with TCC for the

Table 3.4. Production cost for tobacco in Malawi, 2005

Input	Production costs (in MWK)	Percentage of total cost
Seeds	569	1%
Fertilizer	19,582	41%
Other chemicals	370	1%
Manure	635	1%
Materials for barn and sacks	5,623	12%
Annual depreciation and maintenance of tools, oxcarts and oxen	1,004	2%
Club fees	505	1%
Hired transportation/machinery	2,361	5%
Hired labor	14,954	31%
Land rent	135	0.5%
Interest payment	2,046	4.5%
Total input costs	47,784	100%
Net Crop Income	14,315	

Source: Takane 2007.

2011–2012 growing season (Mzale 2012). This jumped to 24,193 registered farmers for the 2012–2013 season as farmers anticipated they would benefit from what turned out to be a temporary increase in tobacco leaf prices.

The auction system

Until the recent arrival of contract farming, there were five main marketing channels for tobacco farmers: auction floors, estates in local areas, farm authorities, unregistered buyers and intermediate buyers (Chirwa 2009). Intermediate buyers include traders with licenses and traders who operate illegally. They purchase tobacco from farmers in urgent need of cash or that are concerned they will not receive a good price at auction. According to Chirwa (2009, 17), of the marketing channels available, "the intermediate buyer channel is more profitable for the farmers. One reason for the profitability associated with intermediate buyers may be a reduction in transaction costs such as transport costs and avoidance of the many levies that are imposed on tobacco when farmers sell directly at the auction or through their affiliated institutions." Approximately 35 percent of smallholder tobacco farmers in Malawi sell their burley tobacco to intermediate buyers (Chirwa 2009).

Tobacco sold to intermediate buyers is generally ungraded and fetches a lower price at auction because leaf buyers want tobacco separated by grade instead of mixed bales with a range of qualities that prompt lower prices from companies. Intermediate buyers encounter problems such as low-quality tobacco due to foreign objects inserted in bales by farmers – sometimes by accident and sometimes on purpose to increase the weight of the bale. Tobacco delivered by unregistered buyers through licensed farmers at auction is associated with relatively high amounts of tobacco containing non-tobacco related materials, and partly accounts for high rejection rates of tobacco sold at auction. Through intermediate buyers or acting alone, some farmers smuggle tobacco to Zambia and Mozambique, oftentimes selling to representatives of local leaf companies with operations in Malawi, to obtain what they believe will be a higher price per kilogram, and to avoid charges at auction.

An estimate of the assembly cost for burley tobacco is presented in Table 3.5. This refers to the expenses incurred at the farm level such as preparation of raw material for market and transportation to auction up to the point of processing by manufacturers. Before farmers arrive at the auction with their product they may be required to store the bales at depots operated by TAMA or the National Smallholder Farmers' Association of Malawi (NASFAM). Farmers incur costs for storage and for transporting tobacco from the depots to auction. Some farmers seek to reduce transportation costs through collectively organized transportation and security at depots. Wait times at storage facilities may be as long as two months. The auction follows a schedule that determines the delivery by district from depots to the market. Transportation costs on average make up almost nine percent of assembly costs (Tchale and Keyser 2010). Farmers that choose not to use depots or are unable to afford depot expenses often store tobacco bales in their living rooms, increasing their risks of respiratory problems due to the inhalation of tobacco dust.

In the next stage of the chain, tobacco is transported to one of the four auction floors in Malawi: Limbe (Southern region), Lilongwe (Central region), Chinkhoma (Central region) and Mzuzu (Northern region). Auction Holdings Limited (AHL), a subsidiary of the state marketing agency Agricultural Development and Marketing Corporation (ADMARC), holds a monopoly on tobacco auction services in Malawi. ADMARC owns 42.63 percent of AHL and ADMARC's chief executive officer (who is appointed by government) is also the chairperson of AHL (Graen 2012). ADMARC's direct involvement in tobacco is limited mostly to the distribution of fertilizer (Chirwa and Dorward 2011). AHL facilitates tobacco auctions and contract marketing.

Farmers' decisions on where to sell the crop are based on distance, congestion (especially at Lilongwe auction floors) and perceived prices (Zant 2012). Distribution of sales among auction floors varies, with Lilongwe being one of the

Table 3.5. Assembly costs for burley tobacco, Malawi

Cost component	Percentage of all costs
Farm production (costs of labor and inputs)	62.1
Purchase from grower (cost of marketing and license)	24.3
Transportation to auction	8.8
Fee, Auction Holding (AHL)	2.4
Fee, Agricultural Research and Extension Trust (ARET)	1.0
Fee, Tobacco Association of Malawi (TAMA)	0.5
Handling charge (TAMA)	0.5
Classification, Tobacco Control Commission of Malawi (TCC)	0.3
Commission, TCC	0.1

Source: Adapted from Tchale and Keyser (2010).

biggest and Chinkhoma the smallest. In 2004, the World Bank's International Development Association funded the development of the Chinkhoma auction floor in an attempt to reduce domestic market costs and decentralize tobacco collection (World Bank 2012; International Development Association and Republic of Malawi 2007). Despite long wait times and relatively poor prices, the Lilongwe auction still appeals to farmers because of its sophisticated technical facilities and its direct connection to the major leaf buyers and their processing factories, which are missing at Chinkhoma and other satellite floors. According to the World Bank (2012, 22), Limbe Leaf and other buyers at satellite floors like Chinkhoma "use their market power to pass on to farmers their added transport costs of auctioning tobacco in more remote areas and transport it to processing centers." At the satellite floors, TCC and AHL do not maintain effective storage facilities or provide reasonable security systems.

After tobacco moves from depots to auction, delivery trucks are required to wait in an area adjacent to the auction until auction authorities indicate that the tobacco may be presented on the auction floors. Farmers have additional expenses for storage while waiting to off-load at the auction, such as food costs paid to truck drivers. Next, tobacco bales are delivered to a holding area near the main floor where they wait for a "runner" with a dolly to retrieve the bale and roll it into one of the rows on the floor. Before auctioning starts, workers from the TCC assign a grade or quality type to each bale. During the auctioning process, buyers licensed by the TCC and auctioneers move quickly through the rows of bales bargaining over the price.

Leaf quality, leaf position on the plant and other details are listed on a tag attached to each bale. Buyers inspect bales that are upright and opened to

Photograph 3.4. An opened tobacco bale rejected due to quality issues (N/S, Auction Holdings)

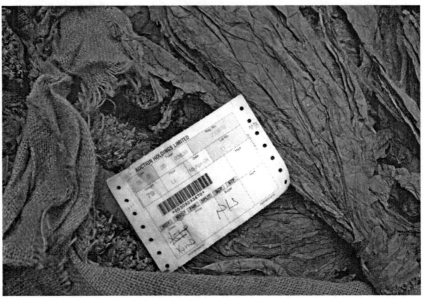

Photo credit: Laura Graen.

determine an offer price. When a bale is sold, the price and buyer initials are marked on the tag. Employees with the auction sew up the bag in preparation for the runner to deliver the bale to a conveyor belt that transports it to processing facilities owned by leaf-buying companies. When a bale is rejected due to quality issues like tobacco being mouldy or foreign objects inserted in the bale, the bale is sent to a re-handling company chosen by the farmer, like Auction Holdings' subsidiary Tobacco Investments Limited or TAMA's Re-Handling Company, and then re-presented on auction. A bale can also be rejected due to disagreements between buyers and sellers over price, in which case it is returned to the holding area and re-presented on the floor later in the season.

After sale at auction, leaf merchants working to fill pre-arranged contracts with cigarette manufacturers take possession of the tobacco and begin processing. Bales are stored according to grade in warehouses operated by Limbe Leaf, Alliance One and other leaf buyers. Workers unpack bales and load leaves onto conveyor belts that lead to threshing machines where stems are separated from leaves. At the beginning of the process, workers wearing masks use their hands to remove inferior quality leaves or any non-tobacco related material. When the process of re-drying, threshing and blending is complete, dried leaves are turned into unmanufactured tobacco that is packed in cardboard boxes and loaded into cargo containers. More than 70 countries receive shipments of Malawi's tobacco, with Belgium, Germany and Denmark being the top destinations in terms of trade value (Business Analytic Center 2008). Cigarette companies may store the tobacco as inventory for up to two years, releasing the product based on need. Cigarette companies such as BAT, which controls 90 percent of Malawi's cigarette market (Shafey et al. 2003), produces cigarettes with local leaf in factories outside of Malawi, selling packets and loose sticks to Malawian adults and children (British Broadcasting Corporation 2008).

The emergence of contract farming

In 2005, TAMA supported the initial development of contract farming in Malawi through the launch of a pilot program with an allocation of five million metric tonnes (Tobacco Association of Malawi 2011). In 2010–2011, the TCC determined that 65 million kilograms (31 percent) of the tobacco produced in Malawi could be sold through the contract system. Allocations were made as follows (Limbe Leaf 2012b):

• Limbe Leaf (18.2 million kilograms)
• Alliance One (16.1 million kilograms)
• Africa Leaf (now JTI Leaf Malawi) (13.7 million kilograms)
• Premium Tobacco (13.4 million kilograms) and
• Malawi Leaf (3.6 million kilograms).

The following season the government and leaf companies agreed that 80 percent of tobacco could be sold through contract marketing (Limbe Leaf 2012e).

Limbe Leaf refers to contract farmers as "business partners" to express the company's commitment to farmers as stakeholders in Malawi's tobacco industry (Limbe Leaf 2012c). Limbe Leaf and Alliance One prefer the contract system to the auction system because companies retain greater control over labor arrangements, yields and prices. Also, the system allows companies to select farmers that have the capacity to uphold industry practices. This increases compliance and traceability, which are requirements cigarette makers place on suppliers such as Limbe Leaf and Alliance One in Malawi (Moyer-Lee and Prowse 2012).

Stancom, a subsidiary leaf company that merged in 2005 with Dimon to form Alliance One International, started contract farming in 2001 with Malawian flue-cured leaf producers. Alliance One administers contracts with 15,000 farmers and Japan Tobacco International (JTI) contracts with 2,800 farmers – a number it plans to increase by 30 percent each year (Jassi 2012).

In the mid-2000s, Limbe Leaf became the largest player in contract farming (Prowse 2011). Details on Limbe Leaf's contract growers are not available, although the company reports that it works with growers to meet a target of 25,000 metric tons annually (Limbe Leaf 2012a). BAT does not directly contract with farmers in Malawi. Rather, it buys four percent of the burley crop in the country through other leaf companies (British American Tobacco 2010). Imperial Tobacco buys about five percent of its tobacco through Alliance One and other leaf companies in Malawi, and does not grow tobacco in the country (Imperial Tobacco 2012).

Through contract burley farming, leaf buyers are reshaping Malawi's tobacco sector and consolidating buyers' power along the marketing chain. In the contract system, farmers have direct growing arrangements with tobacco leaf companies, and prices are determined by the contract that the farmer and buyer sign before the growing season begins. While Moyer-Lee and Prowse (2012) argue that TCC determines prices for contract tobacco, the authors' field observations and interviews were unable to verify TCC's role, if any, in influencing prices for contract tobacco. Contract tobacco is sold through the silent auction system in which leaf buyers have the exclusive right to purchase from their contract growers. Silent auction means that the tobacco bales of contract farmers are presented in a designated area on the auction floor where they are graded by TCC classifiers and then inspected and bought by the contracted company. The system allows the government to collect revenues and facilitates the repayment of credit providers and payments to producer organizations and industrial bodies (Moyer-Lee and Prowse 2012). When the burley contract system began for the 2006–2007 growing season, relatively affluent smallholder farmers with access to capital established contract relationships with leaf buyers (Moyer-Lee and Prowse 2012). During that season, 23.7 percent of total sales volume was of the contract type (Limbe Leaf 2007).

Malawi experienced a sharp drop in tobacco production during the 2011–2012 season, achieving the smallest harvest since 1990. This was a consequence of the devastating 2010–2011 auction season during which rejection rates were high, prices at auction were low and foreign aid cuts provoked the most severe economic crisis in Malawi since democratization. The economic crisis also provoked shortages of fuel and fertilizers. As a result, tobacco farmers lacked money to buy inputs (Graen 2012). In 2011–2012, some 64.63 million kilograms of burley tobacco were sold. Of this total, 29.05 million kilograms (45 percent) were sold through contract and 35.59 million kilograms (55 percent) were sold at auction (Limbe Leaf 2012b). Three companies (Limbe Leaf, Alliance One and Africa Leaf) dominated both auction and contract tobacco trade, accounting for 77 percent and 84 percent of the product purchased, respectively (Table 3.6).

Table 3.6. Leaf buyer market share at tobacco auction and through contract buying, 2007–2012

Leaf buyer	2007		2008*	2009		2010		2011		2012	
Total volume (million kg)	86.6		169.2	208.7		193.2		208.3		64.6	
	Auction	Contract	Contract	Auction	Contract	Auction	Contract	Auction	Contract	Auction	Contract
Limbe Leaf	39%	6%	29% overall	32%	26%	29%	25%	42%	28%	38%	23%
Alliance One	38%	26%	34% overall	36%	26%	25%	24%	32%	27%	27%	32%
Africa Leaf/JTI**	13%	32%	16% overall	12%	23%	13%	25%	10%	23%	12%	29%
Premium TAMA	8%	36%	13% overall	10%	25%	12%	26%	9%	22%	10%	16%
Malawi Leaf	NA	NA	5% overall	8%	NA	17%	NA	6%	0%	13%	1%
ATC	NA	NA	2% overall	2%	0%	3%	NA	0	0	0	NA

* Breakdown by auction and contract type not included in industry reports.
** Japan Tobacco International (JTI) acquired Africa Leaf Processors in 2011.
Source: Data from Limbe Leaf Market Reviews 2006–2012.

In the 2012–2013 growing season, contract farming accounted for 80 percent of total volume trade. This marks the beginning of the end of Malawi's auction system and an increase in companies' vertical power along the tobacco commodity chain. The government of Malawi probably supported the development of the contract system to better match supply and demand of tobacco and reduce price fluctuations. By attempting to increase predictability of the tobacco leaf market the government also likely wanted to achieve an optimal flow of foreign exchange. It means, however, that the government of Malawi has also relinquished control of the leaf-buying system. Government refusal to stop the change from auction to contract arrangements is probably due to pressure from leaf companies to accept the change or face the consequences of a shift of buying to other countries where contract farming is prominent.

The contract system for burley took on a new name at TAMA's 24th Annual Congress in June 2012, which had as its theme "Integrated Tobacco Production System – for Profitability and Sustainability" (Tobacco Association of Malawi 2012). According to Fred Kamvazina, TCC's technical and operations manager, "the IPS [Integrated Production System] was a new concept by name but has been running in the country as contract growing and marketing for the past five years" (quoted by Khanje 2012). Based on pressure from leaf companies dissatisfied with the auction system, the government of Malawi supported IPS (Face of Malawi 2012).

As a tobacco industry initiative, IPS helps prevent over-production since leaf buyers specify quantity and quality requirements with the contract farmers that produce the leaf (Tobacco Journal International 2012). An official with AHL stated that, "the shift [to IPS] may hurt the local economy because some growers will be unable to take up contract farming" (quoted by Jomo 2012). Evans Matabwa of AHL said that under the IPS, not all of the Malawians growing tobacco "will have the opportunity to earn revenue from tobacco, as not all of them will be contracted" (quoted by Jomo 2012).

The IPS is recognized as a system of control that provides tobacco companies with greater influence in the tobacco marketing chain in countries such as Malawi (Prowse 2011) and Brazil (Vargas and Campos 2005). In September 2012, a group of farmers' associations succeeded in launching a court injunction against the formal beginning of IPS in Malawi, arguing that it was anti-competitive (Chinoko 2012). In November 2012, a judge disallowed the injunction and tobacco companies were permitted to continue with the IPS. The case suggests that disagreements between local farmers and leaf companies may become more prominent factors affecting Malawi's tobacco marketing chain. Contract farming weakens farmers' associations. In the auction system, farmers' associations like NASFAM and TAMA provide extension, depot and transport services to farmers. These activities are funded through auction levies. In the contract system, leaf companies provide extension and other services to contract farmers, reducing NASFAM's and TAMA's direct links to farmers (Graen 2012). The rise of the contract system also threatens the very existence of the TCC. As leaf buyers begin to establish their own satellite markets, TCC activities like leaf grading and monitoring of marketing practices become less relevant (Graen 2012).

As the contract system scaled up in 2012–2013, multiple scenarios began to emerge that may have an impact on smallholder and tenant farmers. Farmers may accept exploitative conditions that accompany contract arrangements with leaf companies (as have farmers in other countries with contract tobacco marketing systems). The number of tobacco farmers may decrease as they have limited markets where they can sell tobacco without a contract. Farmers operating outside contract arrangements may have limited access to credit and other resources that contract farmers enjoy. Any benefits of the contract system may be undermined by a lack of competition and the absence of effective regulations in Malawi to prevent permanent debt or to require companies to honor contracts with farmers. Malawi's significant tobacco dependency weakens the government's influence on the leaf companies.

Tobacco prices and the leaf buyer monopoly

Leaf prices, auction and contract sales and revenues earned by leaf companies are key elements in Malawi's tobacco marketing chain. Average prices for burley tobacco sold, and the proportion of total sales at auction and through contract farming, have fluctuated greatly from year to year. Meanwhile, leaf buyer's revenues have increased steadily.

Figure 3.3 shows that auction sales volume and unit values of burley and other tobacco types in Malawi have increased steadily between 1960 and 2010. By contrast, Table 3.7 shows that prices paid to farmers fluctuated from as low as USD 0.91 per kilogram in 2006 to a high of USD 2.39 per kilogram in 2008, dropping again between 2009 and 2011 and only partly recovering in 2012.

Determinants of burley tobacco leaf prices include quality, production volumes and global demand. Determinants of quality and quantity of burley volumes include climate (too little rain, too much rain, late rain, early rain) and the availability of cash or credit to obtain inputs such as fertilizer required to grow tobacco. Production volumes in other burley-producing countries and inventory levels of cigarette manufacturers and leaf merchants influence the global demand for burley tobacco produced in Malawi and other countries.

Figure 3.3. Auction sales volume and unit values of Burley and other tobacco types in Malawi, 1960–2010

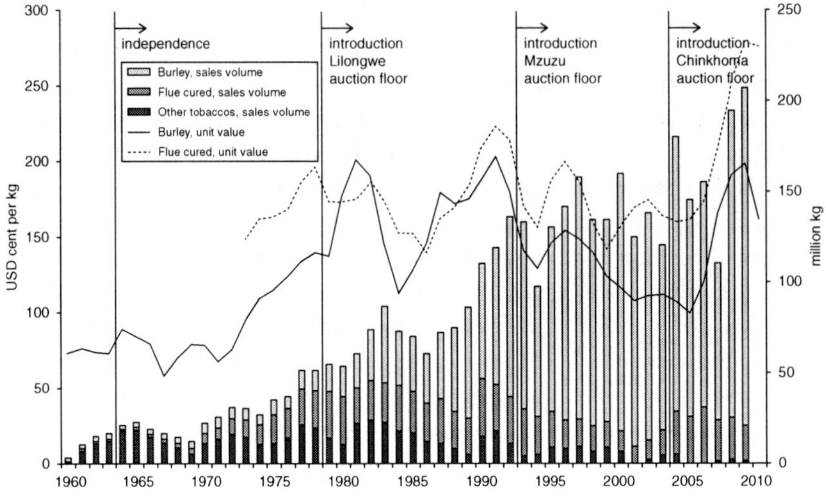

Source: Zant (2012).

Table 3.7. Tobacco auction and contract sales and prices, 2004–2012

Sales and prices	2004	2005	2006	2007	2008	2009	2010	2011	2012
Total contract volumes allocated by TCC (millon kgs)	NA	NA	NA	NA	40	60	40	65	78 Actually sold: 29
Total (millon kgs)	151	120	123	87	169	209	193	208	65
Price per kg (USD)	1.09	0.99	0.91	1.74 (auction)	2.39 (auction)	1.75	1.77	1.13	2.04 (auction)
				1.66 (contract)	2.32 (contract)				2.05 (contract)

Sources: Limbe Leaf Market Reviews (2006–2012); Auction Holdings (2011).

Before the selling season, TCC works with representatives of ARET and TAMA to devise grades for the different types of tobacco produced in Malawi, and according to AHL, each year the industry comes up with a minimum cost of production which is then used for coming up with minimum prices for a season (Auction Holdings 2011). The minimum price allocation was practiced from 2007 to 2011, but abandoned in 2012. The inability of the government to implement minimum prices and tobacco industry pressure on the government of Malawi to support the contract system may have facilitated the end of the minimum price allocation program. Other factors include periodic leaf oversupplies in Malawi and threats from tobacco industry executives to buy leaf at lower costs in neighboring countries.

It is common knowledge in Malawi that the auction stage in the tobacco marketing chain is characterized by non-competitive practices among a few leaf buyers that exert control over tobacco sales. Leaf-buyers' practices are considered monopolistic, with the largest shares in the auction market held by Universal Corporation's Limbe Leaf (38 percent) and Alliance One (27 percent) having relatively large shares of the auction market in 2012. Other companies have smaller market shares: Malawi Leaf (13 percent), JTI Leaf Malawi (12 percent), Premium/TAMA (10 percent) and RWJ Wallace a nominal amount (Prowse 2011). Until the late 2000s, Limbe Leaf and Alliance One had a combined market share of up to 90 percent of leaf purchases. This share was reduced with the development of new leaf buyers in the mid-2000s. One of the key events that created momentum for the increase in the number of leaf buyers was the 2005 Malawi Anti-Corruption Bureau (ACB) report that documented price collusion at auction by Limbe Leaf and Alliance One (formed by the merger of Stancom and Dimon). In a 2012 interview (conducted by Graen), an AHL official confirmed some of the non-competitive practices discussed in the ACB report:

> You'll find that you're seeing bales bought in [an agreed] order, Limbe Leaf, maybe Stancom, then another buyer, another buyer, then after that, Limbe Leaf again. [...] We still sometimes feel that these guys during the night, maybe they have a cup of coffee and say, 'Gentlemen, why not suppress the prices?'

Leaf companies in Malawi are getting better export prices and at the same time suppressing prices in auction and through contract channels. Figure 3.3 and Table 3.6 both show that the difference between contract and auction prices is small, which raises the question: why do leaf companies prefer the contract system to the auction channels historically available in Malawi? Leaf companies have suggested that farmers receive higher household income through the contract system and that companies obtain "long-term supply security" for contracted leaf purchases in Malawi and elsewhere (Alliance One International 2012). It is important to note that companies may exaggerate the price benefits to farmers of the contract system. Leaf companies pursue contract arrangements for a variety of reasons: to reduce uncertainties in the auction system, assert greater control over leaf quality and yields, minimize the purchase of leaf bales with non-tobacco related material that characterize relatively high amounts of leaf bales at auction and meet cigarette manufacturers' requirements of traceability from seed to cigarette (Ryan et al. 2001; Prowse 2011; Moyer-Lee and Prowse 2012). Any price benefit to farmers of contract farming is subject to these company drivers. In June 2012, the president of the Tobacco Association of Malawi (TAMA) stated,

> [A] big step towards dealing with the FCTC is to venture into Integrated Tobacco Production Systems which encompasses measures that deal away with the vices the anti-tobacco lobbyists are capitalizing on, like child labor, traceability issues, good agricultural practices (GAP), etc. Hence this year's timely theme [of TAMA] of "Integrated Tobacco Production Systems – For Profitability and Sustainability" (quoted by Maigwe 2012).

The steady increase in tobacco revenue earned by leaf companies and price differentials at auctions and in contract arrangements with farmers suggest that contract farming does not improve or stabilize prices for farmers. On the contrary, the virtual monopoly of buyers in Malawi provides subsidiary companies of Universal Corporation and Alliance One International with the platform to promote contract farming as the dominant leaf-buying arrangement and further strengthen their direct influence over prices. Meanwhile, at the farm level producers receive low and uncertain prices that are often insufficient to cover the costs of production.

Government Interventions in Malawi's Tobacco Sector

The Malawi government intervenes on multiple levels in the tobacco sector, from cultivation to export. Through the TCC, the Malawi government provides the institutional framework to regulate the tobacco sector. It registers and licenses tobacco growers, leaf-buying companies and auctioneers of tobacco. TCC monitors tobacco production and demand levels, while also forecasting earnings, thus helping the government to plan and meet its budgetary goals. The government owns 42 percent of AHL, the only company permitted to operate auction floors in Malawi. Through compulsory levies at auction, the government also provides financial resources to TAMA, ARET, NASFAM, TCC and AHL. Through its classification services at auction, the TCC provides some protection to farmers from under-grading (the practice of unfairly assigning lower grades to tobacco leaf). This kind of protection is absent in Brazil and other countries that lack government controls and where contract farming is dominant (Vargas and Campos 2005).

Until the start of this century, government involvement in the marketing chain contributed directly to the operation of the tobacco sector. More recently, however, the emergence of contract farming initiatives has sharply eroded the influence of the government on key aspects of the marketing chain, including the management of human resources. Underfunding of the TCC has made the organization vulnerable to poaching of staff by Limbe Leaf and other buyers able to provide more competitive salaries and benefits. In a 2009 interview (conducted by Graen), Henderson Chimoyo, TCC's acting general manager, discussed the problem of TCC-trained leaf graders and classifiers leaving to work for leaf companies:

> [A] good portion of those are trainee classifiers, the really experienced ones we have lost to the industry […] TCC is a government parastatal and our salary structure, our conditions of service are governed by government and looking at our income levels, too, *we haven't been in a position where we could compete with the rest of tobacco industry* where the moment you move out of this office into Limbe Leaf tobacco company, which is a Universal group company or Alliance One International, you're in a totally different area and the conditions are also totally different and it would only take a doubling up of somebody's TCC salary for somebody to change their mind and cross [over to a leaf buyer] (Graen 2012: 93f; emphasis added).

Conflicts that involve poached TCC staff are overshadowed by the direct influence leaf companies have on the TCC. Representatives from Alliance One and Limbe Leaf occupy seats on the TCC board. One TCC board member is Madalitso Mutharika, son of the former President Mutharika (Mponda 2010), who also operates tobacco farms. The makeup of the TCC board itself is exclusionary. Representatives of the Tobacco Tenants and Allied Workers Union of Malawi (TOTAWUM), the union that represents the economic interests of tobacco-related smallholders and farm workers, are excluded from TCC's board.

The TCC is also vulnerable to the power of the executive branch in Malawi. During the 2009 election year, President Mutharika sought to consolidate political power in his Democratic Progressive Party and potentially reduce opposition to his mandated minimum tobacco prices by firing Godfrey Chapola, the TCC general director (Ng'ambi and Banda 2009). Chapola was a supporter of the opposition party and publicly undermined the government's mandated minimum prices when he stated that higher prices were virtually impossible due to the global economic downturn and increasing momentum for stronger tobacco-control policies (Nyasa Times 2009). According to a declassified communication from the US Embassy in Lilongwe dated 22 May 2006 Chapola said privately that Mutharika's prices were too high and that the government would need to back down from its threats to Alliance One and Limbe Leaf (US Embassy in Lilongwe 2006). Chapola's departure illustrates how TCC, as the key government regulatory body in the marketing chain, is subservient to the executive branch in Malawi.

President Joyce Banda is continuing the government's support for corporate interests and development of the tobacco sector. Within two days of emerging as president after the unexpected death of Mutharika in April 2012, President Banda reinvited Charles Graham, the general manager with Limbe Leaf. She also reinvited others who had been forced to leave Malawi in 2009. Her predecessor, Mutharika, had deported tobacco industry leaders who were "stealing" from tobacco farmers by paying low leaf prices and potentially "destabilizing the country" (US Embassy in Lilongwe 2009). Deportees included Charles Graham, Kevin Stainton (CEO of Limbe Leaf), Bertie van der Merwe (leaf-buying manager with Limbe Leaf), Collin Armstrong (managing director of Alliance One) and Alex Mackay of Premium Tobacco (US Embassy in Lilongwe 2009). George Freeman, the CEO of Universal Corporation which is the parent company of Limbe Leaf, applauded this change in tone by President Banda, noting in the company's annual report (Universal Corporation 2012, 4) that, "As an example of our strong local relationships, our senior regional and corporate executives were recently invited to meet with President Joyce Banda, the new president of the Republic of Malawi, to discuss sustainable tobacco production there."

Corporate (Ir)-responsibility and Interference in Government

In the process of supporting tobacco production in Malawi, leaf buyers and global cigarette manufacturers say they are creating a new role for tobacco companies as responsible partners in Malawi's marketing chain. The image enhancement pursued by tobacco companies occurs as part of the broader context in which conflicts over prices are waged among farmers and leaf buyers in Malawi.

Evidence of unscrupulous economic activities in Malawi includes leaf buyers' violation of the US Foreign Corrupt Practices Act (FCPA). According to the US District Court for the District of Columbia (2010, 14), "Between approximately October 2002 and November 2003, Universal Leaf Africa made payments totaling USD 500,000 to one high-ranking Malawian government official; USD 250,000 to a second high-ranking government official; and USD 100,000 to a political opposition leader." Using an account operated by its subsidiary in Belgium, Universal Corporation paid the amounts cited above to secure tobacco contracts in Malawi. Alliance One International also collaborated with Universal Corporation in a coordinated bribery scheme involving tobacco contracts and favors from politicians to influence legislation beneficial to leaf buyers in Malawi, Mozambique, South Africa, Brazil, Thailand, Kyrgyzstan, as well as other countries (US Securities and Exchange Commission 2010). In August 2010, Universal Corporation agreed to pay USD 4.4 million and Alliance One International agreed to pay USD 9.45 million in criminal penalties for conspiring to violate the FCPA and for violating the anti-bribery provisions of the FCPA.

To create a new role and present a favorable image, leaf buyers and cigarette makers currently fund corporate social and environmental responsibility (CSER) schemes. One example is the Eliminate Child Labor in Tobacco Foundation (ECLT) set up by Phillip Morris International, Japan Tobacco International, BAT, Limbe Leaf and Alliance One. Under the project title "Child Labor Elimination Actions for Real Change" partners such as Save the Children Federation of Malawi, Creative Centre for Community Mobilization, Total LandCare and Youth Net and Counseling will deliver activities from 2011 to 2015. Philip Morris, with funding to the ECLT and working in collaboration with Limbe Leaf is sponsoring a 54-month USD 9 million project in Malawi (Limbe Leaf 2012d).

In 2011, Alliance One donated USD 2,900 (MWK 489,300) to the Home of Hope orphanage in Mchinji as part of the company's efforts to create opportunities for vulnerable children and achieve Alliance's CSER goals (Nyasa Times 2011). Other corporate responsibility initiatives funded by Alliance One and other companies strengthen alliances among government officials, farmers' associations such as TAMA and non-governmental organizations that might otherwise criticize or publish details on tobacco industry practices that harm Malawi's farming families. So far, these projects have had a very limited impact on tobacco farmers, and are more correctly seen as a distraction from the more fundamental question of non-competitive practices and unfair leaf prices.

Parallel to CSER activities, alliances among leaf companies, government officials and TAMA actively try to derail tobacco-control policies and circulate claims that tobacco control harms jobs and livelihoods. TAMA is at the center of opposition in Malawi against tobacco control, gaining its power to influence public discussion and debate through its relationship with ITGA, the International Tobacco Growers Association funded by tobacco companies to lobby for the interests of the tobacco companies. TAMA's membership includes tobacco estate growers, tobacco clubs, cooperatives and affluent tobacco growers, with the latter exerting control over TAMA (Graen 2012). TAMA was a founding member of ITGA in 1984. Through TAMA and the ITGA, these elites within the tobacco-farming community obstruct tobacco-control policy making and use Malawi to undermine discussions on the FCTC. For example, Reuben Maigwe (TAMA's president), Felix Mkumba (TAMA's chief executive) and five officials from the TCC and the Ministry of Agriculture actively participated in ITGA's campaign at an international meeting in Uruguay to weaken guidelines to ban ingredients such as sugar and vanilla used in blended cigarettes. These ingredients are known to make cigarettes more addictive (German Cancer Research Center and National Institute for Public Health and the Environment 2012). In a speech welcoming Joyce Banda, President of Malawi, to TAMA's annual meeting on 21 June 2012, Maigwe stated that ITGA represented TAMA directly at the 4th Conference of the Parties (COP 4). He also reported that,

> The Malawi Delegation and ITGA Delegation were always in constant touch and kept sharing strategies to counter whatever was coming from inside the Conference Room. It is pleasing to report that as Observers, the Malawi Delegation together with the ITGA Delegation made an indelible and significant mark to the outcome of the Uruguay discussions to the extent that the banning of burley tobacco never materialized. This shows how serious the Government of Malawi ponders on the welfare of its citizens, more so the tobacco farmers, and let alone the economy of the country. However the battle against the FCTC continues (Maigwe 2012, 3–4).

Four months later in June 2012, Tim Hughes, a consultant with South African-based Read Dillon consultancy group, said during a meeting with key tobacco industry leaders in Malawi that global tobacco-control advocates, through Articles 17 and 18, seek "to stop farmers or curtail farmers from growing tobacco." He described the FCTC process as "highly unfair, irregular, highly undemocratic, high-handed and frankly neo-imperialist if not neo-colonialist to impose what is not a solution onto Malawian farmers without at least consulting with them" (quoted by Campbell 2012).

Graham Kunimba, chief executive officer of TAMA, after attending a day-long conference on the FCTC in Cape Town, South Africa, organized by ITGA and Tobacco Institute of Southern Africa (a group funded by the tobacco industry to lobby on industry agricultural issues), said, "Delegates at the conference wondered why other countries should be speaking for Malawi which is the largest burley tobacco exporter when Malawi could defend her position on some of the FCTC articles" (quoted by Jimu 2012). This seems to reflect TAMA's interest in working to weaken the FCTC from within, rather than a genuine commitment to protect the health and human rights of the population.

In a speech at the annual meeting of TAMA in June 2012, President Banda reinforced Malawi's opposition to the FCTC (Banda 2012, 5).

> I am aware that the "war" on tobacco continues through the Framework Convention on Tobacco Control (FCTC) and that some of the problems currently being faced are a result of the implementation of the convention. Understandably, the FCTC will continue to take its toll by way of having several negative implications on production, marketing and consumption. For sure, we all know that the FCTC has succeeded in reducing the consumption of tobacco in some parts of the world and there has to be a corresponding effect on production for which stakeholders in the industry including Government must stand up to.

Conclusion

In Malawi, companies control the tobacco marketing chain from seed to cigarette, engaging along the way with farmers, farm workers, government authorities and non-governmental organizations funded by the tobacco industry. The companies portray themselves as responsible agents, friends of farm workers and stewards of the environment through corporate social and environmental responsibility schemes. Tobacco companies draw upon this image when arguing that tobacco-control policy making, and not companies' practices, undermine tobacco-related jobs and revenues in Malawi. Closer examination, however, shows that monopoly control of the supply chain allows the industry to routinely buy tobacco leaf produced by children and other unpaid family members, downgrade leaf arbitrarily (by assigning a lower quality), collude among themselves to suppress prices and reduce their costs at the expense of the forest, water table and soil. All the while they make a claim regarding the impact of tobacco-control measures on global demand that has no basis in logic or fact. To more fully substantiate our argument, we believe additional research is needed to show that tobacco farming has been bad for farmers in Malawi and that tobacco farming yields outcomes that are inferior to other types of farming.

Do tobacco-control measures harm leaf prices and earnings? Tobacco companies would like Malawians to think so, and have positioned themselves as defenders of growers. As this paper has shown, virtually all of Malawi's tobacco is sold on the global market, and is consequently insulated from any effects of tobacco-control measures in Malawi. Effects of international efforts to control tobacco consumption are many decades away (see Chaaban, this volume). The real threats to leaf prices and earnings by farmers in Malawi are farmer debt and company practices of price manipulation and increasing control over the production cycle. Unable to obtain decent earnings or fair leaf prices and break free from indebtedness, tobacco families are tethered to the marketing chain. As a result, families send children to fields instead of schools, further deepening their long-term bondage. Meanwhile, through their arguments against tobacco-control measures, tobacco companies deny Malawian policy makers the opportunity to promote the cultivation of food as cash crops or protect its population from the health and ecological costs of tobacco growing.

References

Alliance One International. 2006. "Glossary." Online: www.aointl.com/au/glossary.asp (accessed 12 July 2012).

Alliance One International. 2012. "Annual Report." Online: http://phx.corporate-ir.net/phoenix.zhtml?c=96341&p=irol-reportsannual (accessed 22 December 2012).

Anti-Corruption Bureau. 2005. "Allegation of Corruption and Connivance among Tobacco Buyers, Auction Holdings Limited and Other Stakeholders in the Tobacco Industry." Lilongwe, Anti-Corruption Bureau.

Assunta, M. 2012. "Tobacco Industry's ITGA Fights FCTC Implementation in the Uruguay Negotiations." *Tobacco Control* 21(6): 563–68.

Atwell, W. 2013. "'When We Have Nothing We All Eat Grass': Debt, Donor Dependence and the Food Crisis in Malawi, 2001 to 2003." *Journal of Contemporary African Studies* 31(4): 564–82.

Auction Holdings. 2011. "Tobacco Grades/Minimum Prices." Online: www.ahlmw.com/footer_page_details.php?footer_page_id=10 (accessed 12 August 2012).

Banda, J. 2012. "Speech during the Official Launch of the 24th TAMA Annual Congress in Lilongwe." Online: http://zachimalawi.blogspot.com/2012/06/president-joyce-bandas-verbatim-speech.html (accessed 1 September 2012).

BBC (British Broadcasting Corporation). 2008. "Tobacco Giant Breaks Youth Code." Online: http://news.bbc.co.uk/2/hi/africa/7475259.stm (accessed 19 August 2012).

British American Tobacco. 2010. "Malawi: Child Labor in Tobacco." Online: www.bat.com/group/sites/uk__3mnfen.nsf/vwPagesWebLive/DO86BLLZ?opendocument&SKN=1 (accessed 13 July 2012).

Brown, B. and W. Snell. 2011. "US Tobacco Situation and Outlook." Center for Tobacco Grower Research, Quarterly Newsletter, November 2011.

Business Analytic Center. 2008. *Trends and Prospects in International Trade of Unmanufactured Tobacco; Tobacco Refuse.* Cypress: Business Analytic Center.

Campbell, J. 2012. "Malawi's Economic Conflict: Adhere to the Requests of International Donors or Continue Investing in its Major Cash Crop?" Online: http://jessicamcampbell.wordpress.com/2012/06/13/malawis-economic-conflict-adhere-to-the-requests-of-international-donors-or-continue-investing-in-its-major-cash-crop (accessed 21 November 2012).

Chinoko, M. 2012. "Court Stops Integrated Tobacco Production System." Online: www. zodiakmalawi.com/zbs%20malawi/index.php?option=com_content&view=article &id=5845:court-stops-integrated-tobacco-production-system&catid=42:banner-stories&Itemid=102 (accessed 1 September 2012).

Chirwa, E. 2009. "Farmer Organisations and Profitability in Smallholder Tobacco in Malawi." Working Paper, Economics Department, Chancellor College, University of Malawi. 21 pp.

Chirwa, E. and A. Dorward. 2011. "The Malawi Agricultural Input Subsidy Programme: 2005–06 to 2008–09." *International Journal of Agricultural Sustainability* 9(1): 232–47.

Eriksen, M., J. MacKay and H. Ross. 2012. *The Tobacco Atlas.* American Cancer Society. Atlanta, Georgia. Online: www.tobaccoatlas.org/uploads/Images/PDFs/Tobacco_Atlas_4_entire.pdf (accessed 29 July 2012).

Face of Malawi. 2012. "Stakeholders to Agree IPS Practices." Online: www.faceofmalawi.com/2012/11/stakeholders-to-agree-ips-practices (accessed 6 January 2013.)

FAO (Food and Agriculture Organization). 2003. *Issues in the Global Tobacco Economy: Selected Case Studies.* Rome: Food and Agriculture Organization.

———. 2012. "The Food and Agriculture Organization Corporate Statistical Database (FAOSTAT)." Online: http://faostat3.fao.org/home/index.html (accessed 1 August 2012).

Fisher, M. and P. A. Lewin. 2013. "Household, Community, and Policy Determinants of Food Insecurity in Rural Malawi." *Development Southern Africa* 30(4–5): 451–67.

German Cancer Research Center and National Institute for Public Health and the Environment. 2012. "Additives in Tobacco Products." Online: www.dkfz.de/de/tabakkontrolle/download/PITOC/PITOC_Tobacco_Additives_combined_pdf.pdf (accessed 20 January 2013).

Graen, L. 2012. "Opening Malawi's Tobacco Black Box." Master's Thesis, Department of Anthropology and Philosophy, Martin Luther University Halle-Wittenberg, Germany.

Harrigan, J. 2008. "Food Insecurity, Poverty and the Malawian Starter Pack: Fresh Start or False Start?" *Food Policy* 33: 237–49.

Imperial Tobacco. 2012. "Solutions to Child Labor." Online: www.imperial-tobacco.com/index.asp?page=398 (accessed 24 December 2012).

International Development Association and Republic of Malawi. 2007. "Financing Agreement: First Poverty Reduction Strategy Grant Development Policy Lending." Online: http://tinyurl.com/at7awy2 (accessed 1 September 2012).

Jaffee, S. 2003. "Malawi's Tobacco Sector: Standing on One Strong Leg Is Better than on None." Africa Region Working Paper Series No. 55, The World Bank. Online: www.worldbank.org/afr/wps/wp55.pdf (accessed 25 January 2011).

Jaffee, S. and A. Nucifora. 2005. "Smallholders and Market Linkages: The Ongoing Saga of Malawi Tobacco." Presentation at the workshop linking small-scale producers to markets, Malawi, 15 December.

Jassi, K. 2012. "The Future of the Tobacco Auction Uncertain." *Daily Times* [newspaper]. 4 April.

Jimu, C. 2012. "TAMA Wants Malawi to Be a Signatory to FCTC.'" *Weekend Nation* [newspaper] 10 August. Online: www.mwnation.com/business-news-weekend-nation/8665-tama-wants-malawi-to-be-signatory-to-fctc (accessed 15 September 2012).

Jomo, F. 2012. "Malawi Changes Tobacco-Selling System to Move away from Auctions." Bloomberg, 16 November.

Kadzandira, J., H. Phiri and B. Zakeyo. 2004. "The Perceptions and Views of Smallholder Tobacco Farmers on the State of Play in the Tobacco Sector." Report to the World Bank.

Kaplinsky, R. 2000. "A Handbook for Value Chain Analysis." Report to the International Development Research Centre (IDRC), Canada.

Kasapila, W., and T. S. Mkandawire. 2010. "Drinking and Smoking Habits among College Students in Malawi." *European Journal of Social Sciences* 15(3): 441–48.

Khanje, T. 2012. "Contract Tobacco Farming: Stakeholders to Meet in January for Review." *Daily Times* [newspaper], 25 December.

Koester, U., G. Olney, C. Mataya and T. Chidzanja. 2004. "Status and Prospects of Malawi's Tobacco Industry: A Value Chain Analysis." Report to Ministry of Agriculture and Food Security, Lilongwe, Malawi.

Lea, N. and L. Hanmer. 2009. "Constraints to Growth in Malawi." Policy Research Working Paper for World Bank, Malawi.

Limbe Leaf. 2006. "Malawi 2006 Market Review." Online: http://www.limbeleaf.com/cmreports/Malawi%202006%20Market%20Review.pdf (accessed 25 September 2012).

_____. 2007. "Malawi 2007 Market Review." Online: http://www.limbeleaf.com/cmreports/Malawi%202007%20Market%20Review.pdf (accessed 25 September 2012).

_____. 2008. "Malawi 2008 Market Review." Online: www.limbeleaf.com/cmreports/Malawi%202008%20Market%20Review.pdf (accessed 25 September 2012).

_____. 2009. "Malawi 2009 Market Review." Online: www.limbeleaf.com/cmreports/Malawi%202009%20Market%20Review.pdf (accessed 25 September 2012).

_____. 2011. "Malawi 2010 Market Review." Online: www.limbeleaf.com/cmreports/Malawi%202010%20Market%20Review.pdf (accessed 25 September 2012).

_____. 2012a. "Malawi 2011 Market Review." Online: www.limbeleaf.com/cmreports/Malawi%202011%20Market%20Review.pdf (accessed 25 September 2012).

_____. 2012b. "Malawi 2012 Market Review." Online: www.limbeleaf.com/cmreports/Malawi%202012%20Market%20Review.pdf (accessed 25 September 2012).

_____. 2012c. "Social Responsibility Program Policy Statement." Online: www.limbeleaf.com/srp_policy.html (accessed 25 September 2012).

_____. 2012d. "SRP programmes." Online: www.limbeleaf.com/srp_programmes.html (accessed 25 September 2012).

_____. 2012e. "Proposed Integrated Production System Model for Malawi." Online: http://www.limbeleaf.com/docs/proposed%20integrated%20production%20system%20model%20for%20malawi.pps (accessed 11 April 2014).

Maigwe, R. 2012. "Tama President's Welcome Speech to the Guest of Honour, Her Excellency the State President of the Republic of Malawi, Mrs. Joyce Banda." Tobacco Association of Malawi, Lilongwe, 21 June. Online: http://zachimalawi.blogspot.com/2012/06/tamas-presidents-welcome-speech-to.html (accessed 28 August 2012).

Malawi Government Ministry of Natural Resources, Energy and Environment. 2010. "Malawi State of Environment and Outlook Report: Environment for Sustainable Economic Growth," Environmental Affairs Department: Lilongwe, Malawi. Online: www.nccpmw.org (accessed 12 July 2012).

Mandala, E.C. 2005. *The End of Chidyerano: A History of Food and Everyday Life in Malawi, 1860–2004.* Portsmouth, New Hampshire: Heinemann.

Masanjala, W. 2006. "Crop Liberalization and Poverty Alleviation in Africa: Evidence from Malawi." *Agricultural Economics* 35: 231–40.

Matabwa, E. 2012. "A Stitch in Time: Deleting the Economics of Overdependence." Report to Annual Congress of the Economic Association of Malawi, November.

Mathie, A. and A. Carnozzi. 2005. *Qualitative Research for Tobacco Control: A How-to Introductory Manual for Researchers and Development Practitioners.* Ottawa: International Development Research Centre.

Maulidi, P. 2011. "Malawi Tobacco Revenue Drop by 30%." *Zodiak Malawi* [newspaper] 29 December. Online: http://zodiakmalawi.com/zbs%20malawi/index.php?option= com_content&view=article&id=3911:malawi-tobacco-revenue-drop-by-30 &catid=40:business&Itemid=111 (accessed 11 March 2012).

Mkwara, B. 2010. "The Impact of Tobacco Marketing and Pricing Reforms on Income Inequality amongst Growers in Malawi: What Lessons Can Be Learnt from the Australian Experiences?" Paper presented at the Australian Agricultural and Resource Economic Society National Conference, Adelaide, South Australia, 10–12 February.

Moyer-Lee, J. and M. Prowse. 2012. "How Traceability Is Restructuring Malawi's Tobacco Industry." Working paper prepared for Institute of Development and Policy Management, University of Antwerp, Belgium.

Mponda, J. 2010. "Mutharika Son in Tobacco Control Commission Board: It's Not News Worthy–TCC." *Malawi Voice* [newspaper] 17 November. Online: www.malawivoice. com/business/mutharika-son-in-tcc-board-it%E2%80%99s-not-news-worthy-tcc (accessed 31 May 2011).

Mzale, D. 2012. "Malawi 2012 Tobacco Output Lowest in 18 Years." *Nation Online* [newspaper] November. Online: www.mwnation.com/business-news-the-nation/13761-malawi-2012-tobacco-output-lowest-in-18-yrs (accessed 11 January 2013).

Ng'ambi, M. and M. Banda. 2009. "Bingu Fires TCC Chief." *The Nation on Sunday* [newspaper], 16 March.

Nsiku, N. 2007. "Tobacco Revenue Management: Malawi Case Study." Online: www.iisd. org/pdf/2007/trade_price_case_tobacco.pdf (accessed 4 August 2012).

Nyasa Times [newspaper]. 2009. "Bingu Fires Chapola at TCC." 12 March. Online posting: http://groups.yahoo.com/group/MALAWIANA/message/22121 (accessed 18 August 2012).

_____. 2011. "Alliance One Donates K0.5m to Orphanage." Lilongwe, Malawi.

Organisation for Economic Co-operation and Development. 2012. "Zimbabwe: 2012." African Economic Outlook 2012, Paris, France. Online: www.africaneconomicoutlook. org/en/countries/southern-africa/zimbabwe (accessed 30 September 2012).

Orr, A. 2000. "'Green Gold'?: Burley Tobacco, Smallholder Agriculture, and Poverty Alleviation in Malawi." *World Development* 28(2): 347–63.

Otañez, M., G. Glantz and H. Mamudu. 2009. "Tobacco Companies' Use of Developing Countries' Economic Reliance on Tobacco to Lobby against Global Tobacco Control: The Case of Malawi." *American Journal of Public Health* 99(10): 1759–71.

Otañez, M., H. Mamudu and S. Glantz. 2007. "Global Leaf Companies Control the Tobacco Market in Malawi." *Tobacco Control* 16(4): 261–69.

Otañez, M., M. Muggli, R. Hurt and S. Glantz. 2006. "Eliminating Child Labor in Malawi: A British American Tobacco Corporate Responsibility Project to Sidestep Tobacco Labor Exploitation." *Tobacco Control* 15(3): 224–30.

Prowse, M. 2011. "A Comparative Value Chain Analysis of Burley Tobacco in Malawi, 2003–04 And 2009–10." Working paper prepared for Institute of Development and Policy Management, University of Antwerp, Belgium.

Reserve Bank of Malawi. 2008. "Financial and Economic Review. Volume 40. Number 4." Lilongwe, Malawi.

_____. 2012. "Financial and Economic Review. Volume 44. Number 2." Lilongwe, Malawi.

Ryan, L., F. Gadani, J. Zuber, L. Rossi, C. Archibald and C. Fisher. 2001. "Tobacco Identity Preservation Program: Bates No. 2067576289-6296" Online: http://legacy.library.ucsf.edu/tid/sye34a00 (accessed 15 February 2013).

Shafey, O., S. Dolwick and G. Guindon. 2003. "Tobacco Control Country Profiles 2003." American Cancer Society, Atlanta, Georgia.

Takane, T. 2007. "Gambling with Liberalization: Smallholder Livelihood in Contemporary Rural Malawi." Discussion paper for Institute of Developing Economies. Chiba, Japan.

Tchale, H. and J. Keyser. 2010. "Quantitative Value Chain Analysis: An Application to Malawi." Online: https://openknowledge.worldbank.org/handle/10986/3730 (accessed 27 June 2012).

Tobacco Association of Malawi. 2011. "TAMA's Corporate Profile." Online: www.tamalawi.com/Corporate_Profile.html (accessed 2 July 2012).

_____. 2012. "TAMA President's Welcome Speech to the Guest of Honour, Her Excellency the State President of the Republic of Malawi Mrs Joyce Banda." Online: http://zachimalawi.blogspot.com/2012/06/tamas-presidents-welcome-speech-to.html (accessed 11 January 2013).

Tobacco Journal International. 2012. "TCC Favors Contract Farming." 19 September. Online: www.tobaccojournal.com/TCC_favors_contract_farming.51408.0.html (accessed 14 December 2012).

Tobin, R. and W. Knausenberger. 1998. "Dilemmas of Development: Burley Tobacco, the Environment and Economic Growth in Malawi." *Journal of Southern African Studies* 24(2): 405–24.

Torres, T. 2000. "The Smoking Business: Tobacco Tenants in Malawi." Fafo Institute for Applied Social Sciences, Norway. Online: www.fafo.no/pub/rapp/339/339-web.pdf (accessed 28 July 2012).

UNDP (United Nations Development Program). 2011. "Human Development Report 2011: Sustainability and Equity: A Better Future for All." New York: United Nations Development Program.

_____. 2012. "Africa Human Development Report 2012: Towards a Food Secure Future." New York: Regional Bureau for Africa, United Nations Development Program.

Universal Corporation. 2007. "Supply and Demand Report." Richmond, Virginia.

_____. 2008. "Supply and Demand Report." Richmond, Virginia.

_____. 2012. "2012 Universal Corporation Annual Report." Richmond, Virginia. Online: http://investor.universalcorp.com/common/download/download.cfm?companyid=UVV&fileid=580330&filekey=E215CF66-0477-4BB5-BC0A-F12F6DE06DAA&filename=Universal_Corporation_2012_Annual_Report.pdf (accessed 21 August 2012).

US Department of State. 2011a. "Country reports on human rights practices for 2011." Online: www.state.gov/j/drl/rls/hrrpt/humanrightsreport/index.htm?dlid=186215 (accessed 6 August 2012).

_____. 2011b. "Trafficking in Persons Report." Online: www.state.gov/j/tip/rls/tiprpt/2011/164232.htm (accessed 6 August 2012).

US District Court for the District of Columbia. 2010. "Securities and Exchange Commission v. Universal Corporation, Case 1:10-cv-01318." Washington, DC, 8 August.

US Embassy in Lilongwe. 2006. "Tobacco Market Impasse Risks Forex Crisis." 22 May. Online: http://wikileaks.org/cable/2006/05/06LILONGWE433.html (accessed 11 August 2012).

_____. 2009. "Malawi Deports "Imperialist" Tobacco Executives." 9 September. Online: www.cablegatesearch.net/cable.php?id=09LILONGWE507&q=leaf%20limbe (accessed 11 August 2012).

US Securities and Exchange Commission. 2010. "Litigation Release no. 21618 and Accounting and Auditing Enforcement Release no. 3179." Washington, DC, 6 August.

Vargas, M. and R. Campos. 2005. "Crop Substitution and Diversification Strategies: Empirical Evidence from Selected Brazilian Municipalities." Working paper prepared for International Bank for Reconstruction and Development and World Bank, Washington, DC.

WHO (World Health Organization). 2011. "Malawi Country Profile. WHO Report on the Global Tobacco Epidemic, 2011." Online: www.who.int/tobacco/surveillance/policy/country_profile/mwi.pdf (accessed 1 November 2012).

Wood, B. 2011. "Up in smoke? Tobacco Production's Effect on Childhood Stunting in Malawi." Paper presented at conference of Agricultural and Applied Economics Association, Pittsburgh, Pennsylvania, 24–26 July.

World Bank. 1999. "Curbing the Epidemic: Government and the Economics of Tobacco Control." Online: http://go.worldbank.org/USV7H5C800 (accessed 11 August 2012).

_____. 2005. "Pathways To Greater Efficiency And Growth in the Malawi Tobacco Industry: A Poverty and Social Impact Analysis." Online: http://siteresources.worldbank.org/INTPSIA/Resources/490023-1120841262639/Malawi_Tobacco_PSIA_Synthesis_Paper_April_2005.pdf (accessed 23 November 2012).

_____. 2012. "Implementation Completion and Results Report on Three Credits in the Amount of SDR 68.8 Million. USD 104.7 Million Equivalent to the Republic of Malawi for the Poverty Reduction Support Credits 1-2-3." Online: http://tinyurl.com/atu7skv (accessed 1 September 2012).

Zant, W. 2012. "What Makes Smallholders Move out of Subsistence Farming: Is Access to Cash Crop Markets Going to Do the Trick?" School of Business, Economics and Law in Gothenburg, Sweden, June. Online: www.economics.handels.gu.se/digitalAssets/1373/1373642_zant.pdf (accessed 4 August 2012).

Section Two

TOBACCO-FARMING CONDITIONS IN LOW- AND MIDDLE-INCOME COUNTRIES

Tobacco Industry Myth: Tobacco farmers are currently relatively prosperous and tobacco farming poses no significant risks that cannot be mitigated.

Research Findings:

- Comprehensive cost calculations (that include the cost of unpaid family labor) demonstrate that tobacco farming rarely generates a net gain and often leaves farmers indebted to tobacco companies.
- Tobacco farming is extremely labour intensive and, as a result:
  - Child labour is frequent and leads to missed educational opportunities for children;
  - Women's unpaid labor is dedicated to tobacco farming rather than producing food or independently generating income and resources for the family.

- The tobacco crop generates many unique and serious occupational health hazards, including green tobacco sickness, exposure to exceptionally high levels of toxic agrochemicals and respiratory problems from drying and storing tobacco leaf.
- Tobacco farming in LMICs causes severe environmental damage, including deforestation, land and soil degradation, pollution of waterways and a host of other ecosystem disruptions.

Chapter 4

THE HARSH REALITIES OF TOBACCO FARMING: A REVIEW OF SOCIOECONOMIC, HEALTH AND ENVIRONMENTAL IMPACTS

Natacha Lecours

Multinational tobacco corporations and their associated organizations have for decades claimed that tobacco growing brings prosperity and growth to farmers and farming communities, especially in low- and middle-income countries (LMIC). The industry has also ignored or denied the occupational health and environmental impacts of tobacco farming. Recently, the rhetoric may have shifted. A 2012 study commissioned and funded by the British American Tobacco Company (BATC) concludes that tobacco farming is simply no worse than any other industrial crop (Pain et al. 2012, Part A: 65). It also, based on a sample of only 40 households spread across three countries, argues that there is no evidence of a direct cause–effect relationship between tobacco cultivation and deepening poverty, child labor, indebtedness, food insecurity, environmental degradation or occupational health hazards (Pain et al. 2012, Part B: 74–76). The implication drawn from the study is that the implementation of Articles 17 and 18 of the World Health Organization's Framework Convention on Tobacco Control (FCTC) would force tobacco farmers to turn to less economically beneficial crops for no reason and against their own interests.

This chapter responds to the claims in this study, and more generally by the tobacco industry, that tobacco farming is a good way to make a living and poses no significant risks that can't be mitigated. It does so by examining the available evidence on the economic benefits and costs of tobacco farming in LMICs and the unique health and environmental effects of the industry. The literature that addresses these questions directly is rare and has not been consolidated into a single source. Consequently, our search for relevant information relied on four different channels. First, we identified peer-reviewed articles through a standardized literature search on terms relevant to each impact area using electronic databases (Academic Search Complete, CAB abstracts, GEOBASE, Google Scholar, Medline, SciELO and Scopus). Second, we used Google to identify books, NGOs' reports, videos, etc., documenting aspects of the tobacco industry. Third, we searched the International Development Research Centre's (IDRC) public and internal databases for research reports and documents by researchers funded by the centre. Finally, we reviewed the reference lists of all the identified relevant documents to find additional sources. Visits by the author to IDRC-funded research projects in Bangladesh, Cambodia, Kenya, Malawi and Vietnam, and review of the case studies presented in this book also contributed to the synthesis of evidence under each impact area presented below. The findings clearly show that tobacco growing is not simply just another industrial cash crop but rather a crop that undermines the sustainability of farming communities and imposes both private and public costs detrimental to the short- and long-term prosperity of smallholder farmers and economic growth in LMICs.

Economic and Social Impacts

The literature on tobacco farming in LMICs overwhelmingly concludes that smallholder tobacco farmers are struggling. Financial costs often outweigh benefits due to the intensive use of labor in the various stages of tobacco production (including family labor), the high cost and large quantity of external inputs such as fertilizers and pesticides used to support the crop, variable and uncertain tobacco yields, local manipulation of tobacco leaf prices by tobacco traders and the financial burden of recurring indebtedness. Evidence regarding each set of problems, described below, is a cause for concern by governments and merits further attention from the research community, civil society organizations and relevant government sectors (such as ministries and agencies in charge of agriculture, labor, rural development, health, etc.).

Poor returns to labor

It is well known that tobacco farming requires much more labor than most crops. While the length of the tobacco-growing season varies from country to country, the plant needs a high level of care, wherever it is grown. Often with limited tools and technology, farmers need to open up the fields, establish and tend nurseries, top and spray the plants, harvest the leaves progressively as they ripen and cure and grade the leaves before sending or taking them to the market (Sejjaaka 2004). In places where fuel wood is scarce, farmers report having to walk 10 to 17 km to fetch wood from the forests, adding significantly to their labor costs (Muwanga-Bayego 1994; Waluye 1994).

Farmers often complain about the amount of labor needed for tobacco farming compared to the economic gains resulting from it. In Cambodia, farmers said that, "After adding up all the costs related to tobacco production (including paying wood for drying tobacco leaves and labor) and the fluctuation of tobacco prices in the market, [...] the benefit from tobacco production is less than that from other crop production (Bunnak et al. 2009, 21)." They also noted that the profit from tobacco growing was not proportional to the amount of labor required and the costs of its production. Samrech's 2008 study of tobacco growing in Cambodia, which concluded that tobacco farming remains the most profitable crop for farmers in the country, is flawed in this respect because it does not include labor costs in the calculations of profitability. The study does not consider the actual returns to labor or the opportunity costs faced by tobacco-farming households.

A Vietnamese study (Nguyen Thanh et al. 2009) points to the impact of labor costs on overall profitability. For example, it found that the labor costs of planting 1,000 square meters of tobacco was approximately twice as high as the labor costs of planting 1,000 square meters of maize or rice in the studied communities. Bunnak et al. (2009) mention that tobacco farmers in Cambodia also raised similar problems of high labor costs and low net returns. A Lebanese study (Hamade, this volume, 56) found that, "Tobacco farming is unprofitable when labor costs are factored in, and cannot be sustained at the level of small-scale production without being subsidized." In China, a study found that when measuring the ratio between the costs and revenue for each crop, tobacco had a lower return than oilseeds, beans and fruit (Hu et al. 2007). Detailed cost–benefit analyses of tobacco farming in Bangladesh (Akhter et al., this volume), Kenya (Kibwage, this volume) and Brazil (Almeida, this volume) also show that returns to labor in tobacco are much lower than for other comparable crops.

Other studies have assessed household assets and livelihood strategies among both tobacco-growing and non-tobacco-growing households, and established that the annual net incomes of non-tobacco farmers were higher than those of tobacco farmers (Kibwage et al. 2009). Vargas et al. (2009) have also observed that farmers (of the Rio Pardo Valley region in Brazil) that do not have tobacco as their main cultivar own more durable goods than tobacco growers. In general terms, the Rio Pardo Valley region presents, along with high levels of tobacco production, socioeconomic development indexes that are lower than those of other municipalities in the state of Rio Grande do Sul (Almeida, this volume).

In regards to opportunity costs, some studies have found that non-tobacco farmers have more time to engage in other income-generating activities (such as multi-cropping, fishing, trading and day labor) than tobacco farmers (Espino et al. 2009; Guedes de Lima 2007). Studies have also shown that tobacco farming limits women's ability to engage in other income-generating activities that used to be part of their livelihood, such as trade or food cultivation (Arcury and Quandt 2006; Babalola 1993; Babalola and Dennis 1988). Sejjaaka (2004) argues that the failure to value family labor imposes an additional injustice on African women who are drawn by their husbands into tobacco farming when their real interests lie in producing food crops. When obliged through family power relations to help in tobacco farming and curing, the time they can spend on food crops, domestic chores and caring for children is reduced. As discussed below, this increases food and economic insecurity for families and other problems in the household.

The different ways used to measure profitability and economic returns to tobacco make it difficult to generalize for LMIC farmers. Babalola (1993) and Heald (1991), for example, focus on the socioeconomic status of tobacco farmers and their ability to generate capital and make investments when discussing the profitability of tobacco farming. It could be that in some contexts, farmers who engage in tobacco-growing activities are already better off economically, either from owning larger plots of land (which allows them to cultivate both subsistence food and cash crops) or from having more resources to begin with. In Cambodia, the director of a district's Agriculture Department stated that, "Farmers who grow tobacco are those with better standard [sic] of living. They own one or two pairs of oxen, ox cart and other means of production. [...] Poor farmers cannot farm tobacco (Bunnak et al. 2009, 16)." While this may be true in some contexts, in others poor farmers do farm tobacco on small plots of land or rent land under contract. They too may rely entirely or almost entirely on their earnings from tobacco to buy food and other household necessities (Akhter et al. 2008; Nguyen Thanh et al. 2009; Sejjaaka 2004).

It is also important to consider which crops are being compared to tobacco when assessing profitability. Some traditional food crops, like corn and black beans in Brazil, provide a much lower net income than tobacco (Vargas and Campos 2005). However, these foods are not usually grown for cash, but for subsistence. The costs of tobacco production are about five times higher than the costs of corn production and six times higher than the costs of black bean production. Discussing crop substitution and diversification strategies in Brazil, Vargas and Campos (2005) have noted that in the Rio Pardo Valley, where tobacco growing is widespread, some alternative food crops (a combination of agroecological products in one case and bananas in another) have yielded higher net incomes than those earned from tobacco.

Besides the costs of labor and fuel wood, the costs of owning or renting land have also often been discounted when assessing tobacco's profitability. As noted by Akhter et al. (this volume), a very significant number of tobacco farmers in Bangladesh are tenants paying high rents for access to suitable land. Moreover, studies are starting to show that tobacco farmers must spend a larger proportion of their income on healthcare services (Hoang Van et al. 2010; Kibwage et al. 2009; Samrech 2008) as a result of the occupational hazards of tobacco growing, a figure that is also often left out of the profitability equation.

Dependency and debt

Tobacco industry representatives applaud themselves for providing services to smallholder farmers that may not be readily available from other sources. Research shows, however, that the services – mainly access to inputs and market infrastructure and technical assistance – come at a stiff price in dependency and debt. This price is not always obvious to farmers, as inputs are provided at no cost at the start of the season and deducted from the final lump-sum payment at the end of the season. The practice of advancing inputs, a central feature of contract farming, often results in farmers being left with debt at the end of the season, either to tobacco companies or intermediary traders. This in turn prompts a return to tobacco the following year in an attempt to pay off the previous years' debt. Choice in crop, and scope for transition to other farming livelihoods, is severely limited, perpetuating the heavy work burden borne by all members of the household, including women and children.

Kirk (1987, 46) defines contract farming as "a way of organizing agricultural production whereby small farmers or out-growers are contracted by a central agency to supply produce in accordance with conditions specified in a contract or agreement. The agency purchasing the produce may supply technical advice, credit and other inputs, and undertakes processing and marketing." The practice of contract farming, he observes, emerged in the post-colonial period when "the removal of the political shield afforded by the colonial authorities led to changes in the ownership and control of land (1987, 46)." In this new context, it became increasingly difficult for transnational corporations (TNCs) to directly control production of agricultural products through their own plantations. Unionization of the labor force, improved communications and economic diversification also undermined the political and economic viability of the plantation model. Faced with rising labor costs, challenges in land ownership and issues around worker recruitment and control, TNCs, including tobacco companies, have increasingly turned to contract farming practices.

In comparison to production on plantations, contract farming reduces the costs and risk of investment in the following ways (Kirk 1987). First, the contracting agency does not need to own the land. This limits capital investment in the land and costs related to land management. Second, because farmers are not directly employed under contract arrangements, contract farming provides a means to avoid labor-related problems and disputes. It also allows the contracting agency to reduce expenditures on labor management and supervision. Third, although production costs and risks are transferred to the farmers, contracting agencies still exercise considerable control over agricultural production. They achieve this by controlling the supply of credit and inputs and by retaining control over the processing and marketing of the product. Moreover, although contracts specify prices and grades, these are usually determined by the purchasing agency. Farmers have little room for negotiation, especially where the agency is a monopsony. Fourth, as opposed to plantation production, and by operating through smallholders and subsidiary companies, contracting agencies acquire a positive image with governments that struggle to support farm livelihoods. Fifth, contracting with farmers enables TNCs to share investment risks with development agencies, financial institutions and local governments, which also often consider contract farming in a positive light. In fact, many governments in LMICs welcome contract farming as a way to not only attract foreign finance and expertise and provide export earnings but also to incorporate smallholder farmers in the national economy without drawing on government revenues and services.

Contract farming is very common in tobacco-producing LMICs, typically involving legal agreements between smallholder farmers and large, transnational tobacco corporations (Vargas and Campos 2005). In these contracts farmers commit to follow the technical guidance of the firm and provide it with tobacco leaves according to a price classification scheme set by the firm. This means that farmers are bound by the volume, quality and production costs defined by the firm. The tobacco corporation assumes responsibility for providing farmers with seeds and technical advice, selling them the inputs (fertilizers, pesticides), advancing and controlling loans, providing transportation and buying the crop. Contract farming thus allows firms to control the specific variety of each tobacco species (Burley, Virginia, etc.), production targets and production costs. This dependency and high level of external control create very asymmetric bargaining powers between smallholder farmers and tobacco TNCs. It is interesting to note that the same dynamic is also observed in China, where the government controls the production of tobacco leaf and cigarettes through its two national monopolies, The State Tobacco Monopoly Administration (STMA) and the China National Tobacco Company (CNTC). Like transnational corporations,

CNTC signs contracts with farmers that specify the amount of acreage under tobacco cultivation and the price for different grades of leaf quality (Hu et al. 2007). Although CNTC is run by the State, Hu et al. (2007) note that the monopoly status of the company allows it to use its own grade quality benchmark, which puts farmers in a weak position for bargaining for higher prices and can lead to downgrading.

Sejjaaka (2004) reports that in 2001 British American Tobacco Uganda used only about 15 percent of its gross income to cover the costs of buying dried tobacco leaf from farmers. This was equivalent to paying USD 150 to each of its 65,000 farmers, an income well below Uganda's per capita income of USD 216 that year (Sejjaaka 2004). The low cost of acquiring inputs to cigarette manufacturing accounts in large part for the high profitability of the industry.

Research from many tobacco-producing countries points to the process of grading the quality of tobacco leaves as a mechanism through which tobacco TNCs forcibly reduce their costs. Tobacco farmers in Uganda, Kenya, Malawi and Bangladesh commonly protest that they are being intentionally cheated by systematic under-grading of their tobacco leaf (Sejjaaka 2004; Kibwage et al., this volume; Otañez and Graen, this volume; Akhter et al., this volume). Farmers in Vietnam also believe that under-grading is common and that prices set in advance by the company are excessively low (Nguyen Thanh et al. 2009). They said that the company takes advantage of their urgent need for cash and the fact that they must sell their tobacco no matter what the price is, or else the tobacco will rot and they will lose everything. In 2009, many farmers in Bangladesh also had difficulty selling their tobacco leaf due to claims by the companies that (despite their control over-production targets) there was a production surplus (Akhter et al., this volume). The farmers felt cheated when the tobacco companies greatly reduced the price per kilogram they were promised at the start of the season and significantly undergraded the tobacco leaves.

Studies in Kenya, Vietnam and Bangladesh have also found many examples of recurrent indebtedness and economic insecurity among smallholder tobacco farmers. As one Vietnamese farmer explains, "Many people who grew tobacco could not sell their products, so they were indebted because they could not pay for fertilizers and other things already bought in advance (Nguyen Thanh et al. 2009, 33)." In Bangladesh, Akhter et al. (this volume) state that an increasing number of tobacco farmers believe they cannot switch from tobacco to other crops because they are indebted to tobacco companies and moneylenders. Since farmers receive a card upon contracting with the British American Tobacco Company (BATC), contracted farmers in Bangladesh are known as "card holders." As the authors explain, the relationship of indebtedness

Photograph 4.1. Farmers in Kushtia (Bangladesh) discussing their sense of betrayal over a lower price than promised in their contract

Photo credit: Abdul Zabbar.

among contracted farmers is reinforced by leasing arrangements between landowners and tenant farmers:

> Intermediaries lease-in land from owners for extended periods of time and lease-out land and agricultural inputs to land-poor farmers. Many intermediaries are also card holders that sell agricultural inputs and buy tobacco products from smaller farmers for resale to the BATC. This set of relationships ties small, medium, and large farmers together in direct and indirect contracts with tobacco companies. The contracts in turn severely limit the extent to which farmers at any scale can make independent land-use decisions or negotiate on prices for inputs and products. Social tensions and conflicts due to this dependency are particularly severe in indigenous communities where land use decisions have traditionally been made collectively through community leaders (Akhter et al. 2008, 13).

Sejjaaka (2004) also concludes that indebtedness robs smallholder farmers of meaningful choices. His research with farmers in Uganda shows that they would like to diversify their crops and earnings by growing coffee and maize but feel they have no choice but to turn to tobacco. Since the problem of farmers' indebtedness is embedded in complex social relations, research is needed to determine how the cycle of indebtedness could be broken and how indigenous knowledge can be mobilized toward creative solutions (see Akhter et al., this volume).

Intra-household inequities

Scarce documentation of the gender impacts of tobacco farming exists and most of it has become dated. Still, literature on the negative gender impacts in the African context points to potential recurrent problems on the continent as well as in other LMICs. For example, in discussing the returns to women's tobacco-farming labor in Nigeria, Babalola and Dennis (1988) observed that tobacco production created gender-based conflicts in Igboho (a Yoruba community in Oyo State). They explain that Yoruba society usually consists of large settlements in which the division of labor is specialized according to male and female tasks. Women are responsible for taking care of the children, as well as providing materially for them through food production. They also have labor obligations to their husbands, upon demand. Along with distinct labor roles, a separation of male and female incomes also exists at the household level. Since the Yoruba are by and large polygamous, wives find themselves competing for the resources (land and capital) controlled by their husband.

In this context, women engaged in tobacco farming were not able to dedicate time to traditional ways of generating income such as small-scale retail marketing and craft occupations such as weaving, dyeing and pottery. Since the introduction of tobacco contract farming, women have been obliged to provide significant amounts of time and labor to their husband, which diminishes their chance of engaging in other income-generating activities and providing materially for their children. This situation was exacerbated by the fact that the Nigerian Tobacco Company was the only leaf buyer in the region. Since the company only contracted with male heads of families, the payments for tobacco bales went entirely to them, giving them complete control over the proceeds from tobacco (Babalola 1993). This created new tensions between husband and wives, as it challenged the traditional norms of labor and income divisions within households and made women even more economically subordinate to their husbands. Similar cases have been reported among the Teso (Heald 1991) and Kuria (Arcury and Quandt 2006) ethnic groups in Kenya.

More recent studies also show that women tend to be responsible for more of the tobacco work than men, due to a division of labor based on gender. With the help of children, women engage in watering tobacco seedlings, transplanting, fertilizing, topping and suckering, harvesting, sorting, stringing and grading the tobacco leaves (Babalola 1993). Although men are usually responsible for the tasks of hoeing and weeding the fields, women also help with these (Babalola and Dennis 1988). In sum, women do most of the work, with some help from children and men. In a Vietnamese study, most respondents agreed that women's work accounted for more of the labor, 60 to 70 percent of the total amount of tobacco production activities (Nguyen Thanh et al. 2009). The study, along with others in Africa (Babalola and Dennis 1988; Muwanga-Bayego 1994; Sejjaaka 2004; Arcury and Quandt 2006), stress that tobacco-farming work puts an extra burden on women, who also have to bear and rear children, manage the household and provide food and other necessities for the household.

Photograph 4.2. Woman farmer in Kenya sorting, tying and grading tobacco leaves for sale

Photo credit: Sandy Campbell, IDRC.

Some studies have reported that tobacco farming encourages multiple marriages, although this is also a cultural practice in many countries. Two main reasons have been given for this. First, some studies found that farmers are encouraged to have several wives to supplement family labor on their tobacco farms (Kibwage et al., this volume; Kibwage et al. 2009; Arcury and Quandt 2006; Sejjaaka 2004). Second, other studies have noted that men use the extra cash income provided by the sale of tobacco to attract other wives. In Uganda, Sejjaaka (2004) noted that some farmers only grow tobacco to pay the dowry for a bride, and then stop growing it. In Kenya, Heald (1991) observed that Teso women feared that a husband would use tobacco income to marry a second wife, which would mean that they would have to compete with another woman for the products of joint labor and for the land that would later be divided among her sons. During visits by the author to tobacco-farming communities in Kenya in 2010, male farmers confirmed that multiple marriages were explained by both of these reasons.

In addition, the problem of spending sprees by male tobacco farmers has been widely reported, in newspaper articles and through informal discussions during field visits by the author in Kenya, Malawi and Bangladesh. The problem stems from men receiving lump-sum payments at the end of the tobacco-growing year (see Akhter et al., this volume). Men engage in spending sprees and use money earned from tobacco to engage in substance abuse and prostitution. In the African context, the traditional division of income at the household level may partly explain why men may not feel responsible for using tobacco income to provide for basic household needs or share with their wives. Further research is needed to better understand men's perception of their role as economic provider within their household, as well as other possible explanations for this behavior (such as a false sense of wealth due to lump-sum payments, poor financial management skills, etc.).

Children also suffer from the extra burden that tobacco production imposes on households. Although it is common practice for children to contribute to family agricultural work in LMICs, studies have shown that the labor intensiveness of tobacco farming worsens the situation and contributes to the highest rates of child labor across sectors (Otañez et al. 2006; Nguyen Thanh et al. 2009; Arcury and Quandt 2006). Otañez et al. (2006) reported that an estimated USD 10 million per year is contributed in unpaid child labor in Malawi, a country highly dependent on tobacco growing. The International Labor Rights Forum (ILRF 2008) reported that in 2008 an estimated 78,000 children worked on Malawian tobacco plantations. Research on the working conditions of tobacco child laborers by Plan Malawi (2009) found that children worked long hours for little pay (which was also regularly withheld) and suffered from physical and sexual abuse from their supervisors on tobacco farms.

A study by Amigó (2010) also found that child labor represented a substantial proportion of very low-paid tobacco-farming labor in Indonesia. Men were paid substantially more than women and children for an equal number of worked hours (40 percent more than women and 120 percent more than children). According to the author (2010, 39), "These pay differentials are not a matter of labor supply and demand because there appears to be a shortage of all types of labor during the tobacco season [with men performing tasks such as hoeing and compressing the leaves, and women and children performing other time-consuming tasks]. Instead, they reflect the subaltern position of women and children in Indonesian society."

The negative effects of the extra work burden on children is especially felt in seasons of peak labor demand. At these times children are forced to miss school so that they can work on tobacco tasks. In Malawi, a country highly dependent on tobacco, "most of the children drop out of school completely, and when they are about 16 years old they themselves may start to grow tobacco on land they rent from a land owner (ILRF 2008, 10)." Kibwage et al. (this volume) also found that tobacco-farming households invested less income on children's education compared to non-tobacco-farming households in similar conditions. This reduces children's future access to the labor market, which could offer a way out of tobacco growing. The children are also exposed to the nicotine and pesticide poisoning related to tobacco work. Their potentially higher vulnerability to these effects has yet to be studied, notably across important grower countries like China, Brazil and India, among others (Mcknight and Spiller 2005; Arcury and Quandt 2006). This gap in knowledge is especially important in light of the tobacco industry's efforts to deflect attention away from the harm of tobacco growing to children by offering inadequate corporate social responsibility (CSR) program responses. As an example, Otañez and Graen (this volume) explain that five transnational tobacco companies set up the Eliminating Child Labour in Tobacco Growing Foundation (ECLT) as part of their corporate social responsibility agendas. As research results on the industry's CSR activities tend to show, Otañez et al. (2006) found that the modest efforts undertaken by British American Tobacco in Malawi through the ECLT fund (building schools and wells and planting trees) were designed to distract public attention away from the business model that allowed them to profit from children's low wages and unpaid work as opposed to taking meaningful steps to eradicate child labor in the Malawian tobacco sector.

Knowledge gaps

While several recent studies have identified many of the same problems in different LMICs, most of the peer-reviewed literature dates from the 1980s and

1990s and is limited to the African region. Therefore, a need exists to deepen the documentation of the socioeconomic problems associated with tobacco growing in Africa today, as well as in Asia and Latin America. Addressing national information needs is important, as are contributions to the body of peer-reviewed literature that can more easily be accessed by professionals working in the fields of tobacco control and international development.

In general, more information is needed to better understand five important socioeconomic impacts:

1. The cost–benefit ratio of tobacco farming, with calculations taking all production costs into account (labor, fuel wood, land rental, debt and land depreciation).
2. The extra work burden imposed on women. For example, more information is needed on the amount of time women spend on all tobacco production activities.
3. Ethnographic studies analyzing how different labor and household income divisions affect gender relations. Outside Africa, little is known about women's loss of economic autonomy due to tobacco growing.
4. The extent of child labor in tobacco production and its effect on educational achievements and children's health.
5. Debt cycles and broader social dynamics associated with contract farming practices.

Health Impacts

Four serious health risks are prominent in tobacco-growing communities: green tobacco sickness (GTS), exposure to agrochemicals, respiratory diseases and food insecurity due to the displacement of food crops. GTS (caused by the dermal absorption of nicotine) and some respiratory diseases are unique to the tobacco sector and affect a significant proportion of farmers, leading to a host of recurrent short-term symptoms and potentially unknown long-term and chronic impacts on health. Available evidence on agrochemical exposure, which is responsible for skin and respiratory disorders as well as poor neurological and psychological health, points to increased health risks for tobacco farmers due to high levels of exposure in tobacco farming. In addition to these medical conditions, some tobacco-growing countries were found to face increased food insecurity due to the displacement of food crops. The devotion of land to tobacco at the regional scale is leaving both tobacco and non-tobacco-growing communities vulnerable to unstable and uncertain markets for food. While most peer-reviewed research on the prevalence of these risks and their health effects among tobacco farmers and workers are from the United States, research in LMICs is beginning to validate the severity of these concerns.

Green tobacco sickness

The most researched health problem associated with tobacco cultivation is green tobacco sickness. GTS is "nicotine poisoning that results from the absorption of nicotine through the skin from contact with tobacco plants during cultivation and harvesting (Arcury and Quandt 2006, 72)." Because nicotine is an alkaloid that is water and lipid soluble, it dissolves in water on the leaves of the green tobacco plant. Direct skin contact with green tobacco leaves, or contact with the water from these leaves, results in exposure and absorption of nicotine (Arcury and Quandt 2006).

McBride et al. (1998) compiled what had been learned about this disease in studies conducted between 1966 and 1998, mainly among American tobacco harvesters. Since then, Arcury and Quandt (2006), Schmitt et al. (2007) and Riquinho and Hennington (2012) have conducted literature reviews on the health impacts of tobacco production and harvesting. These studies found evidence mainly from the United States, with only a few examples from LMICs.

The studies show that working in wet conditions (wet tobacco and wet clothes) increases the risk of exposure to nicotine and development of GTS (Arcury and Quandt 2006). Water, rain, dew or perspiration can draw nicotine out of plants and thus facilitate its absorption through the skin. Consequently, the geographical clustering of GTS cases is influenced by rainfall, temperature and humidity (McBride et al. 1998). The kinds of tasks performed on the farm also influence exposure. Among all farm workers, pickers were found to have higher nicotine levels and more GTS symptoms (McBride et al. 1998; Arcury and Quandt 2006).

Studies also show that the different types of tobacco grown and the way tobacco is harvested produce different exposures. GTS is primarily found among workers who hand-harvest and handle tobacco leaves before curing (McBride et al. 1998). Flue-cured tobacco, which is harvested progressively as the leaves ripen, presents even greater opportunity for exposure than burley tobacco, which is harvested as a whole stalk (Schmitt et al. 2007). In the former case, the whole body is exposed to the plant as the harvester picks the tobacco leaves. In the case of shade-grown tobacco (which is grown under cloth, to protect it from direct exposure to sunlight), skin exposure is limited. The leaves of shade-grown tobacco are usually picked when they are dry, further reducing potential nicotine absorption (Schmitt et al. 2007).

There are no established diagnostic criteria for GTS, but the common symptoms are dizziness, headache, nausea, vomiting, pallor, weakness, increased perspiration and chills. Symptoms can also include abdominal cramps and pain, prostration, difficulty breathing, diarrhea and fluctuations in blood pressure or heart rate (McBride et al. 1998; Arcury and Quandt 2006). Although different in other ways, GTS symptoms are similar to organophosphate poisoning (a type of pesticide commonly used in tobacco cultivation) and heat exhaustion. This can cause misdiagnoses where health professionals are not aware of GTS as a problem faced by tobacco farm workers (McBride et al. 1998).

GTS is normally self-limiting and of short duration, but symptoms may be severe enough to result in dehydration and the need for emergency medical care or hospitalization (Arcury and Quandt 2006). Treatment usually involves suggestions to stop work that requires contact with green tobacco, to wash

Photograph 4.3. Boy harvesting tobacco leaves in São Lourenço do Sul, Rio Grande do Sul

Photo credit: G. E. G. Almeida.

and change clothes, as well as encouragements to increase fluid intake, ingest dimenhydrinate (to treat nausea and vomiting) and rest (McBride et al. 1998).

Arcury and Quandt (2006) provide reliable estimates of prevalence and incidence of GTS through a study conducted in two North Carolina counties. They concluded that the overall prevalence for GTS was 24.2 percent (44 out of 182 individuals) and that the overall incidence density was 1.88 (for every 100 days of work in tobacco, tobacco workers had GTS for 1.88 days), with the incidence density varying across the tobacco production season. A review of GTS studies by Schmitt et al. (2007) shows a very wide range of findings, from as low as 8 percent prevalence to 89 percent. They suggest that the highest reported prevalence may be due to unspecific case definitions. For example, Ghosh et al. (1979) defined GTS as any neurological and/or respiratory complaint among tobacco farm workers. They agree with Acrury and Quandt (2006), however, that findings on incidence density across the different studies show a consistent figure of about two cases per 100 days of work (Schmitt et al. 2007).

Very few studies have looked at tobacco farmers' health beliefs and their knowledge of GTS. The few that did found that 50 percent of the farm owners in the United States correctly suspected nicotine to be the cause of GTS, while the other 50 percent "attributed GTS to the posture during harvest, the heat, the smell of the tobacco plants or a combination of these factors (Schmitt et al. 2007, 258)." Farm owners also seemed to have a strong belief that tobacco harvesters developed a tolerance to GTS after working in tobacco fields for some time. These authors concluded that farm owners tended to underestimate the duration of illness, and discounted the seriousness of GTS, in an attempt "to discount their responsibilities in terms of provision of prevention strategies for their employees (2007, 260)."

By contrast, farm workers (who do most of the handling and harvesting of leaves) overwhelmingly believed that GTS symptoms were caused by the application of chemicals to the tobacco plant. Unlike farm owners, however, those who did recognize the real cause of GTS symptoms were also aware that people did not develop tolerance to exposure. Nevertheless, the majority of farm workers and farm owners believed that using tobacco, and being used to working in tobacco, protected them from GTS symptoms (Schmitt et al. 2007; Rao et al. 2002). Interviews with farm owners by Arcury et al. (2003) found that some farm owners believed that the mostly non-smoking Latino farm workers should start smoking tobacco to decrease GTS symptoms. The literature reviews that examined the issue (Arcury and Quandt 2006; Schmitt et al. 2007) found no consensus in the research community regarding the extent to which the use of tobacco products has an impact on GTS symptoms.

It is clear, however, that the use of protective, water-resistant clothing and chemical-resistant gloves reduces the amount of nicotine absorbed by workers in contact with green tobacco plants (McBride et al. 1998; Schmitt et al. 2007). McBride et al. (1998) suggested that the use of protective gear, which is not required by occupational health regulations in most places, should be encouraged. However, the authors warn that these actions should be weighed against the increased risk of heat stress caused by wearing impermeable clothing in hot weather.

Arcury and Quandt (2006), in their review of the literature and own research have concluded that work experience, as indicated by age, also seems to be protective for GTS. According to the authors, this finding "may indicate self-selection (those most sensitive may not return to work in tobacco in subsequent years) or learning better protective measures (2006, 74)."

More research is needed to confidently establish estimates of the prevalence of GTS outside the United States, and factors affecting uptake of protective measures. Studies to date are nevertheless worrisome. A recent study conducted in Vietnam found a GTS prevalence rate of 39 percent (Hoang Van et al. 2010), which is higher than the prevalence rate observed by Arcury and Quandt (2006) for tobacco workers in the United States. This difference may be due to underestimates in the US studies, which rely on hospital-treated cases only, without considering undeclared or untreated cases captured through detailed interviews in the Vietnamese case.

The Vietnamese study also arrived at statistically significant results showing that females and people older than 45 years had a relatively greater number of GTS episodes. Similarly, a study conducted in Brazil found that older subjects presented higher rates of GTS (Almeida 2008). These results contradict with the findings of the United States' studies discussed above, which have identified age and working experience as protective factors against GTS (Arcury and Quandt 2006). It is possible that in the case of Brazil and Vietnam, where older farmers were shown to be more susceptible to GTS symptoms, protection from GTS by leaving the sector may not be possible. While migrant farm workers in the United States might choose not to return to work in tobacco farms, Brazilian and Vietnamese farmers may lack economic alternatives that allow them to stop farming tobacco. Nor do they have the same scope to wear protective clothing as workers in the United States, due to the cost and much hotter, humid working conditions. This explanation of the difference between the two populations is consistent with findings by Riquiho and Hennington (2012), who have looked at obstacles to protecting people from pesticide exposure in Brazil. Researchers concur that more attention should be given to engaging with frontline healthcare workers in areas where tobacco is grown to raise awareness about the symptoms of GTS and document its prevalence (McBride et al. 1998; Schmitt et al. 2007; Hoang Van Minh et al. 2010).

Agrochemical exposure

Exposure to agrochemicals is a common problem among farmers growing a wide variety of industrial crops, due to the reliance on chemical methods of pest control. Tobacco growing is however of particular concern because it routinely uses much higher amounts of chemicals than most other industrial crops (Tobacco Free Kids 2001; Arcury and Quandt 2006; Akhter et al., this volume). These chemical products include insecticides, herbicides, fungicides and fumigants as well as growth inhibitors and ripening agents (Arcury and Quandt 2006). Tobacco farming also involves the application of chemicals at many more stages in the production cycle than grains or even vegetables, increasing the risk of exposure. While in the United States and other high-income countries tobacco tractors pull the chemical sprayers, in LMICs hand-held and backpack sprayers are the common practice. This practice further exacerbates the risk of exposure. In short, the total volume of applied pesticides is high, they are applied at various times throughout the production cycle and in LMICs tobacco farmers use higher risk equipment.

While there are few studies on the health effects of agrochemical use specifically focused on tobacco, their conclusions are consistent with the concerns raised in other sectors. Lonsway et al. (1997) found that both mixing and spraying of chemicals for tobacco fields in the United States led to chemical exposure, each representing a greater risk for workers in relation to the type of chemical used (acephate exposure was greater during mixing while methamidophos exposure was greater during spraying). Other studies have also shown that even tobacco harvesters not directly involved in mixing and applying pesticides are exposed to agrochemicals and run the risk of pesticide poisoning. The findings of Panemangalore et al. (1999) show that tobacco farmers' exposure to pesticides, growth regulators, and/or nicotine reduced the activity of blood and plasma enzymes, a finding confirmed in vitro. While there are a limited number of comparable medical studies in LMICs, the studies presented below show that tobacco farmers are exposed to a variety of very toxic agrochemicals and experience significant symptoms from their exposure.

In Kenya, tobacco farmers showed acetylcholinesterase inhibition during periods of pesticide exposure, which is an indicator of organophosphate and carbamate poisoning (Ohayo-Mitoko et al. 1997). In Malaysia, one-third of 103 tobacco workers presented two or three symptoms related to pesticide toxicity (Cornwall et al. 1995). Another study (Kimura et al. 2005) showed that Malaysian tobacco farmers handling pesticide experienced peripheral and central nervous systems symptoms attributable to chemical toxicity. In Southern Brazil, Salvi et al. (2003) noted an unexpectedly high number of symptoms of the Parkinson type among tobacco workers, as well as anxiety disorders, major depression and suicidal tendencies. While more studies are needed to better understand neuropsychiatric conditions in tobacco farmers, Arcury and Quandt (2006, 75) state that the "accumulating evidence of a link between organophosphate exposure and psychiatric diagnoses (depression and suicidal tendencies) among agriculturalists supports these allegations of psychiatric pesticide hazards among tobacco workers."

In terms of protection, hand washing does significantly remove pesticides from the skin (a 23 to 96 percent reduction, depending on washing method, solvents and time between exposure and decontamination). However, even in high-income countries (HICs) tobacco farmers seem to be unaware of health risks associated with pesticide exposure, or lack the necessary information or resources to protect themselves effectively from it.

In Greece, Damalas et al. (2006a) looked at the main health beliefs related to pesticide use among tobacco farmers. Despite a high level of awareness (99 percent) of the potential health risks posed by pesticide handling, 46 percent of the sampled farmers reported not using any protective equipment when spraying pesticides. The main reason given was that it was uncomfortable (68 percent). Other reasons included that the equipment was too expensive to buy (17 percent), time-consuming to use (8 percent), not available when needed (6 percent) and not necessary (2 percent). Another study conducted

Photograph 4.4. Farmer in Kushtia (Bangladesh) applying pesticide to tobacco plants while children weed the field

Photo credit: Abdul Zabbar.

by Damalas et al. (2006b) found that the labels on pesticide containers did not communicate safety measures effectively to users. More specifically, the study showed that 72 percent of the surveyed farmers found the information on pesticide labels hard to read, while 94 percent found it hard to understand. Perhaps because of that, only six percent of farmers indicated that they paid attention to safety precautions, environmental hazards and first aid and antidote information found on the label, whereas 46 percent stated that they normally exceeded the recommended rates indicated on the labels. In light of these results, the authors suggested that although farmers had adequate knowledge of the potential hazards of pesticide use on health, farmers were not yet adopting safer work practices (Damalas et al. 2006b).

Interviews in 2010 by the author in Bangladesh and Cambodia provide some insight into additional sources of risk to tobacco farmers in LMICs. In Bangladesh, researchers identified 47 different kinds of pesticides used among tobacco farmers in their study area (Akhter 2010). Farmers interviewed by the author complained that they needed to use large amounts of chemicals in tobacco farming and that this had led to pollution of waterways and the degradation of soil, along with individual and community health impacts. In Cambodia, researchers noted that pesticides used by tobacco farmers were often smuggled across borders with no package information or with information in the wrong language. The routine use of persistent dangerous pesticides (DDT) and non-persistent pesticides that have been banned in HICs[1] is reinforced by their ready availability and the general lack of labels and information on how to use them with minimal risks.

Respiratory diseases

Respiratory diseases among tobacco farmers and workers are linked to exposure to high levels of tobacco dust during tobacco processing (Arcury and Quandt 2006). These post-harvest activities include curing (drying of tobacco leaves, often with wood smoke), baling (compacting leaves into bales) and sheeting (tying tobacco into burlap sheets).

The medical community has described a condition known as "tobacco worker's lung," which is a parenchymal lung disease (known as exogenous allergic alveolitis or hypersensitivity pneumonitis) that is caused by the inhalation of tobacco moulds (Nefedov et al. 1991; Olade and Lessnau 2006).

---

1   Arcury and Quandt (2006) report that, according to US Customs' documents, nearly 65 million pounds of banned or restricted pesticides were exported from the United States between 1997 and 2000. Most of these were shipped to less-regulated LMICs, adding to what are already significant supplies of non-patented, acutely toxic chemicals manufactured and easily available in these countries.

The review of literature in the United States by Schmitt et al. (2007) found evidence that tobacco farm workers are also likely to be at increased risk of suffering from disorders of the upper airways such as nasal dysfunction. Studies in Zimbabwe (Osim et al. 1998) and India (Ghosh et al. 1979) also found significantly lower lung function among tobacco workers compared to the general population, with an increased risk of emphysema.

Other evidence points to potential inhalation or ingestion of cadmium, a toxic metal hazard high in the tobacco sector (Dowla et al. 1996). Extracted from the soil by the tobacco plant and sequestered in the leaves, cadmium can then be inhaled by farmers during tobacco-farming work. The authors found that the toxic metal inhibited blood enzymes in a way similar to pesticide intoxication.

In most LMICs, flue-cured Virginia tobacco is the most widely grown variety. This leaf variety must be cured in airtight barns, at a constant heat level (for 72 to 96 hours), to dry properly. During this period, men and women have to maintain the fire and consequently breathe in large quantities of smoke. In Bangladesh, women are usually responsible for performing this task, often without interruption for several days. Common symptoms are chest pains, strain and fatigue, due to bad air quality, stress, lack of sleep and inappropriate nutrition. Women caring for children often experience extra strain and children also suffer from exposure to these health hazards (Akhter et al. 2008; Akhter et al., this volume).

Farmers in Kenya have also raised a number of concerns related to respiratory problems experienced during the curing and storage phases of tobacco production. The authors' interviews with farmers in the South Nyanza Region of Kenya determined that curing takes six to eight weeks. Farmers (mainly women) enter the curing barn frequently to add leaves and branches to an open fire and in the process inhale large quantities of smoke. Farmers interviewed complained of regular chest pains during these months.

Respiratory problems are also caused by storing tobacco in closed spaces, including the home. Farmers reported this practice and associated health problems in Malawi (ICRISAT 2009), Kenya (pers. comm.) and countries across Southeast Asia (SEATCA 2008). The reasons for storing cured tobacco leaves indoors are many. First, after tobacco is cured, farmers have to grade it according to quality, package it in bales and store it until it is transported to the market or auction house. In most areas where tobacco is grown the heavy rainy season starts right after the tobacco harvest, forcing farmers to undertake these activities indoors. Second, smallholder tobacco farmers typically do not own special storage rooms for their dried tobacco. The only space available is their house. Finally, farmers interviewed stressed that they stored the dried tobacco inside the house, and preferably next to their bed, to avoid losing the product to theft. Tobacco farmers using this practice also report that the smell of tobacco leaves and the inhalation of fine tobacco dust are responsible for respiratory problems as well as dizziness, nausea and headaches among their family members. While further research is needed to substantiate the prevalence of respiratory diseases among tobacco farmers and the various causes of these conditions, it is evident from the experience to date that the sector presents unique and serious dangers to respiratory health.

Food insecurity

The impact of tobacco farming on food insecurity seems to be largely a function of farm size. Evidence from Kenya (Kibwage et al. 2008), Brazil (Vargas et al. 2009) and Nigeria (Babalola 1993) suggests that tobacco farmers in these countries tend to have larger plots of land and are able to grow food crops alongside the tobacco crop. Evidence from Malawi (Tobin & Knausenberger 1998), Vietnam (Nguyen Thanh et al. 2009) and Bangladesh (Akhter et al., this volume), however, suggests that tobacco farmers in these countries (other than estate farmers) typically have very small plots of land and dedicate all of their available land to tobacco production. This limits their ability to grow staple foods and tobacco simultaneously and may increase their food insecurity. In the words of a Vietnamese farmer, "Growing rice will directly give everyday food for the families but growing tobacco means we can get starved to death if we can not sell the tobacco. Last year, we already experienced this problem (Nguyen Thanh et al. 2009, 33)." In Africa, as described in the section above, the traditional labor and income divisions between husband and wife, as well as the intensity of tobacco-related labor imposed on women, decrease the amount of food crops grown (Babalola and Dennis 1988; Heald 1991).

Although food insecurity at the household level varies with different socioeconomic status, studies suggest that widespread tobacco farming can significantly reduce the availability of food at the regional level. According to Kibwage et al. (2009), since the 1970s, the land under tobacco in Kenya grew in acreage at the direct expense of food crops. In the South Nyanza region, traditional crops like cassava, millet and sweet potatoes – important in periods of drought and famine – are now scarce and livestock production has fallen drastically. In Bangladesh, tobacco production has displaced food and other economic crops from prime agricultural lands (Akhter et al., this volume). For example, the district of Kushtia, the second largest tobacco-producing district in the country, used to be a food surplus region but is now food insecure. Tobacco has not only displaced crops such as pulses, sugar cane, jute and vegetables, but also made it difficult to transition back to food crops due to the loss of soil nutrients and the build-up of persistent weeds. These authors argue that tobacco production is a threat to the country's food security, and have called for laws that would limit the expansion of tobacco production into new agricultural lands.

Knowledge gaps

Although most research on the health impacts of tobacco farming focuses on farm workers in the United States, the findings and their limitations provide a basis for guiding research on how these effects are experienced in LMICs. GTS studies have mainly focused on white tobacco farmers, although Arcury et al. (2003) and Trapé-Cardoso et al. (2005) have worked with Latino migrant and seasonal farm workers. McBride et al. (1998) says that GTS is underestimated in the US because it relies on hospital-treated cases only, without considering undeclared or untreated cases. The imprecise estimate of GTS prevalence and incidence that could be caused by the different case definitions found across studies is another important limitation in the US studies. The long-term effects of chronic skin absorption of nicotine also need to be investigated in order to better understand impacts on the nervous and cardiovascular systems (Arcury and Quandt 2006). Addressing these knowledge gaps and methodological limitations are important as the great majority of tobacco farmers and farm workers are now in developing countries and are often among the poorest of the poor.

Detailed documentation of the different types and quantities of agrochemicals used in tobacco farming, as well as how they are applied (with or without protective measures), would also help to clarify the unique levels of risk posed by the tobacco sector in LMICs compared to other industrial crops in the same country or region. Research is also needed to determine the neurotoxic effect of pesticide exposure and its relationship to mental health.

Reviews of tobacco literature from the United States by Arcury and Quandt (2006) and Schmitt et al. (2007) have identified other types of health problems associated with tobacco-farming work that have not been studied in LMICs. Research on accidents at tobacco farms in Kentucky showed that falls in tobacco curing barns were the most frequent cause of injury, with the majority of accidents resulting in broken bones. Work in high barns also puts workers underneath at risk from falling objects from levels above (such as sharp, pointed sticks). The lifting of sticks bearing heavy loads of green leaves also resulted in repetitive motion injuries. Research documenting the kinds of working conditions in curing barns in LMICs would help establish what other occupational hazards associated with tobacco exist in these countries.

Schmitt et al. (2007) found a meta-analysis of case-control studies on female tobacco farm workers in Europe that revealed a significantly higher risk of bladder cancer in this population. However, the researchers have not adjusted their analysis according to pesticide exposure, which is potentially a confounder since it is also linked to increased cancer risk. The long-term effects of both agrochemical and GTS exposure must be further investigated to determine whether they can lead to malignancies or other chronic conditions. The authors also synthesized the results of studies looking at different skin disorders. While the number of studies on this subject is not sufficient to offer solid evidence, contact eczema and urticaria were observed in tobacco-farming contexts and seem to be triggered by direct skin contact with tobacco leaves (Nakamura 1984; Szarmachz and Poniecka 1973).

Finally, studies in Kenya (Arcury and Quandt 2006; Kibwage et al. 2009) have pointed to increased reproductive health risks for women involved in tobacco cultivation, but the links between the two remain unclear. During the author's project visits in Bangladesh, female farmers mentioned that they believed tobacco-farming activities to be responsible for more miscarriages among women in their communities. Research is still needed to better understand not only women's reproductive health in relation to tobacco-farming activities, but women's and children's health in general as they perform farming tasks that expose them to severe physical and emotional strain, as well as occupational risks such as skin nicotine absorption, agrochemical exposure and respiratory diseases. Emerging research from Bangladesh (Akhter, pers. comm.) suggests that tobacco-growing households spend more of their income on medical fees than non-tobacco-growing households, which points to tobacco farming as being an inherently hazardous occupation in LMIC farming communities.

Environmental Impacts

Tobacco farming in LMICs is one of the most environmentally destructive monocrops. First, the plant's susceptibility to pests and diseases requires intensive use of agrochemicals. Tobacco also requires the application of fungicides and herbicides to curb the growth of persistent weeds typically found in tobacco fields. The plant also extracts nutrients from the soil more rapidly than many other crops – a response exacerbated by the deliberate practice of topping and suckering to promote the concentration of nicotine in the plant's leaves. In addition to degrading soils very rapidly, the plant residue does not offer any opportunities for soil replenishment. To continue farming tobacco on the same lands, large amounts of fertilizers are needed after the first few seasons.

Second, curing tobacco leaves requires large amounts of fuel wood, which is not needed for other commercial cash crops. When the demand for fuel is met with wood, this provokes deforestation, and when it is met with straw and crop residues, it provokes the loss of soil organic matter and long-term soil degradation. The demand for high levels of soil fertility to support the tobacco crop also stimulates land clearing, which adds to the burden of deforestation. The literature reviewed below illustrates two main environmental impacts: deforestation and pollution due to the heavy use of agrochemicals. Unfortunately, efforts by tobacco companies to mitigate environmental impacts through reforestation and soil amendment practices, when they occur at all, routinely fail to address even a small portion of the damage done.

Deforestation

Virginia tobacco, one of the varieties most widely grown worldwide, requires flue-curing. Flue-curing is the process by which tobacco leaves are dried or cured by means of heat transmitted through pipes or flues, without exposure to smoke or fumes. In high-income countries (HICs), flue-curing is usually done through highly-specialized equipment, which limits inefficient use of energy. In LMICs, flue-curing is mostly done in mud kilns or barns, by burning large quantities of wood. Other varieties of tobacco grown in LMICs are fire- or smoke-cured, where tobacco is cured over open fires in curing barns and in direct contact with the smoke. Both curing methods used in LMICs are highly inefficient in terms of energy use and require large quantities of fuel wood, which tobacco farmers acquire from their own land, surrounding forests, public lands or markets. Often, fuel wood resources are scarce or difficult to access.

In recent decades, tobacco growing gradually shifted from HICs to LMICs. From the 1960s onward, Geist et al. (2009) observed two trends in the global production of commercial tobacco. On the one hand, the share of tobacco produced in the developing world has steadily increased, moving from 57 percent in 1961 to 86 percent in 2006. This represents a 180 percent increase. In 1961, 70 percent of the world's land devoted to tobacco was found in LMICs – a percentage that rose to 90 percent in 2006. This represents a 47 percent increase since 1961. On the other hand, tobacco production has decreased in HICs, falling from 1.5 million tons in 1961 to 0.9 million tons in 2005. The same is true for land under tobacco, which went from 1.2 million hectares in 1961 to 0.4 million hectares in 2005.

Inevitably, the shift of tobacco production from HICs to LMICs has resulted in significantly reduced production costs for transnational tobacco companies.

It has also meant that tobacco-growing LMICs have borne the environmental and social costs of increased tobacco production in their countries. In the 1980s, the issue of deforestation in tobacco-growing countries was already a serious problem, and many organizations started to raise concerns, including the Food and Agriculture Organization (FAO) and the World Health Organization (WHO) of the United Nations (Geist 1999; Chapman 1994). Unfortunately, since the international community lacked, at that time, the scientific data to expose precisely the extent of deforestation caused by tobacco farming, numbers were based on questionable estimates.

In a 1994 *Tobacco Control* journal editorial entitled "Tobacco and Deforestation in the Developing World," Chapman (1994) summarized the popular claims on this subject. The summary reported that in 1976 Muller claimed that one tree was required to cure every 300 cigarettes, a statement repeated in a WHO publication (WHO 1980). In 1993, Madeley suggested that 12 percent of all world deforestation was caused by tobacco curing, an enormous figure now discounted. Elsewhere, authors claimed that trees from one hectare of land were needed to cure a hectare of tobacco, while others estimated twice that ratio.

In response to claims regarding tobacco-led deforestation, the tobacco industry commissioned a report to evaluate its impact on global deforestation. Known as the International Forest Sciences Consultancy (IFSC) report, it was published in 1986 and authored by A.I. Fraser. The report was commissioned by the International Tobacco Information Centre (INFOTAB), which was funded by the tobacco industry. The report examined the fuel wood consumption of Argentina, Brazil, Kenya, Malawi, Zimbabwe, India and Thailand, then extrapolated the data to 69 tobacco-growing developing countries (Chapman 1994). Adjusting scales to allow for comparison, the IFSC report used the Specific Fuel Consumption (SFC) index, which refers to the number of kilograms of wood required to cure one kilogram of tobacco. According to the report's calculations, made to assess the popular estimations of fuel wood use, the "one tree for 300 cigarettes" claim would equal 230 kg of wood per kg of tobacco, and the "one hectare of wood for each half hectare of tobacco" would equal 100 kg of wood per kg of tobacco. In contrast, the IFSC report states that "the average SFC found in 300 barns in the seven countries studied was a remarkably low 7.8 kg per kg, with a range of 2.5–40 kg per kg among the farms (Fraser 1986, cited in Chapman 1994, 192)."

Unfortunately, following the IFSC report, no independent research to assess the global level of tobacco-related deforestation was done until 1999 (13 years later). Nonetheless, Geist (1999) was able to demonstrate the significant importance of tobacco production as a cause for global deforestation.

His results also suggested that the impacts of tobacco-related deforestation were felt more significantly on certain producer countries and regions in the developing world.

> The average amount of natural vegetation removed per developing country is more than 2000 ha or about 5% of total national deforestation, while it rises, on average, to around a quarter of all deforestation in the group of seriously affected producers. As a major factor contributing to crop-specific deforestation, the global mean of flue-cured produce using wood is only about 12%, but increases to a mean 62% in the producer countries with minor-to-serious tobacco-related deforestation (Geist 1999, 25).

Moreover, the author highlighted that high deforestation rates are especially threatening to the fragile drylands and uplands environments in which tobacco is grown. Drylands cover 30 percent of the world's surface and accommodate a large proportion of the world's poorest people. Mainly caused by large-scale deforestation for agricultural purposes, an estimated 70 percent of global drylands are affected by desertification due to land degradation. Upland areas are also prone to accelerated deforestation, since they provide favorable conditions for agriculture compared to lowland and humid environments.

One of the regions highly impacted by tobacco-related deforestation is the Southern African region covered with Miombo woodlands.[2] The impacts on the forest ecosystem of that region were first examined in the 1990s (Waluye 1994), and more recently through several studies conducted in Tanzania (Yanda 2010; Abdallah et al. 2007; Abdallah and Monela 2007; Sauer and Abdallah 2007; Mangora 2005). Generally, these studies confirm that there is serious tobacco-related deforestation in the region, as well as forest and soil degradation. In Kenya, tobacco-related environmental problems documented in the 1990s (Kweyuh 1994; Waluye 1994; Muwanga-Bayego 1994) were still found to be present in 2009, including widespread deforestation and the felling of indigenous trees for curing, as well as soil erosion, change of local streams from permanent to seasonal and water pollution from agrochemicals used in tobacco production (Kibwage et al. 2009).

---

2   As described by Abdallah and Monela (2007), the appellation "*Miombo* woodlands" comes from a local word that ecologists adopted to describe the woodland ecosystems dominated by trees of the genera *Brachystegia*, *Julbernardia* and *Isoberlinia* (*Leguminosae*, sub-family *Caesalpinioideae*).

Another important cause of tobacco-related deforestation that has emerged from this literature is land clearing. Several of the studies mentioned above (Yanda 2010; Mangora 2005; Abdallah et al. 2007; Sauer and Abdallah 2007) discussed the agricultural practice of shifting cultivation, and the serious threat it poses to the sustainable use of the Miombo woodlands. In fact, smallholder farmers in Tanzania frequently obtain their tobacco plots by clearing forest land through shifting cultivation. According to Abdallah et al., "Shifting cultivation is, by far, the leading land-use change associated with nearly all deforestation cases (96 percent)," making small-scale subsistence farming in the region one of the major threats to forests (2007, 93).

Sauer and Abdallah have argued that "tobacco production in Tanzania is still dominated by small-scale subsistence farmers highly dependent on family labor, hand tools, natural resources, as well as animal-drawn farming implements (2007, 422)." Because more technical inputs are beyond the reach of most small-scale tobacco growers, the expansion of their production mainly happens through more land clearing. Mangora (2005) also looked at the social and cultural reason for shifting cultivation,

Photograph 4.5. Wood collected from the hillsides of the Chittagong Hill Tracts (Bangladesh) and floated down the Matamuhuri River feeds the kilns for tobacco grown along the river bank

Photo credit: Abdul Zabbar.

which remains the major farming system in Urambo District (Tanzania). Virgin land is preferred for tobacco growing because farmers fear soil-borne diseases and expect a higher yield from it. Mangora found that in the district, 69 percent of tobacco farmers cleared new woodlands for tobacco cultivation each season, while only 25 percent of them grew tobacco on the same lot for two consecutive seasons, and only six percent did so for more than two consecutive seasons. Consequently, the fallow periods became as low as four years, as opposed to an original fallow time of ten years. According to the author, such significantly shortened fallow periods threaten the recovery capacity of the woodlands and will eventually cause a change of land-cover from woodlands to bushlands or lead to permanent deforestation. Finally, because tobacco cultivation is dominated by small-scale farming, which highly depends on forest resources for acquiring new arable land and for the curing of the crop, all the Tanzanian studies cited above conclude that tobacco farming is not sustainable in the way it has been and is still practiced in the region.

The relentless demand for fuel wood by the tobacco industry also leads to indirect economic hardships. For example, farmers in Cambodia reported that rubber trees (used as a livelihood source) were being cut around the communities for curing tobacco (Bunnak et al. 2009). In parts of Bangladesh where fuel wood is scarce, tobacco farmers use fodder, rice straw and fruit trees to cure tobacco. These practices then affect food supply and resources (cooking fuel and food for milk cows) and overall food security (Akhter et al., this volume).

Soil and water degradation due to agrochemical use

While many forms of industrial farming pollute the water and degrade the soil with various kinds of chemicals, tobacco has particular attributes that add to this general environmental burden (Tobacco Free Kids 2001; Arcury and Quandt 2006; Akhter et al., this volume). Particularly vulnerable to pests and diseases when grown in a monocrop, tobacco needs the addition of fungicides and herbicides to curb the growth of diseases and persistent weeds typically found in tobacco fields. The plant also extracts nitrogen, phosphorus and potassium from the soil more rapidly than many other crops – a problem exacerbated by practices such as topping and suckering that promote the concentration of nicotine in the plant's leaves (Tobacco Free Kids 2001; Geist et al. 2009). In addition to depleting soils very rapidly, the plant residue does not offer any opportunities for soil replenishment. To continue farming tobacco on the same lands, large amounts of fertilizers are needed after the first few seasons.

Research results from Brazil and Bangladesh have highlighted a number of long-term environmental problems caused by agrochemical use. In Bangladesh (Akhter et al. 2008; Akhter et al., this volume), research has shown that the use of chemicals to control a persistent weed found in tobacco fields (commonly known as "mula") is polluting the water, killing fish and destroying soil organisms that are needed to maintain soil health. In addition, unlike food crops, tobacco production offers no return to the soil and the ecosystem. Since the biomass (stalks or plant residue) left after the harvest offers no ecological or economic value to the farmers (it cannot be eaten by animals or used as fuel), farmers report a loss of livestock and poultry in their households. In turn, the diminished animal population translates into a loss of animal manure, essential to maintain soil health in developing countries.

In Brazil, a number of studies have identified excessive chloroform, phosphorus and agrochemical residues in waterways adjacent to tobacco-farming communities (Gonçalves et al. 2005; Griza et al. 2008; Bortoluzzi et al. 2006). In these cases, water pollution was caused by agrochemical residues, and exacerbated by reduced forest cover around the communities, which helped to transfer other pollutants to the water. The monitoring of a catchment area in Southern Brazil concluded that the shift to more intensive tobacco production in ecologically fragile areas, such as wetlands, riparian zones and steep slopes, resulted in severe impacts on hydrological systems and sediment yield (Merten and Minella 2006).

Finally, it is important to note that the tobacco industry is promoting agricultural practices with many negative environmental effects (Lecours et al. 2012). First, through its control of the leaf production system, the industry promotes and contributes to the sale of large quantities of agrochemicals, which are harmful to environmental and human health. Second, documentation exists to illustrate how tobacco companies constantly shift their production operations from degraded to fertile environments, which actively contributes to resource mining (Akhter et al., this volume). Examples from Bangladesh, Honduras, Brazil and Kenya describe how big transnational corporations, especially British American Tobacco (BAT), have exploited regions for a period of time before moving out completely or in part, and exploiting new and lush regions to fulfill their supply needs at the lowest possible cost. Third, the literature shows well that, by investing in corporate social responsibility (CSR) campaigns and activities, the industry exacerbates the problems by addressing them inappropriately – and often by not addressing them at all – whilst trying to create a positive image of their business in the realms of politics and public opinion (Chapman 1994; Tobacco Free Kids 2001; Akhter et al., this volume). As reported by Lecours et al. (2012) the inadequacy of the industry's CSR activities in relation to deforestation is well illustrated by four African studies (Kweyuh 1994; Waluye 1994; Muwanga-Bayego 1994; Kibwage et al. 2009). In sum, the authors have highlighted that reforestation initiatives promoted fast-growing exotic trees such as cypress and eucalyptus. These replacement species were inappropriate as they required extra care to grow and survive in the African environment and extracted large quantities of groundwater, which was not the case of indigenous species. This approach was clearly designed to rapidly serve the fuel wood needs of the industry, as opposed to restoring indigenous ecosystems that do not impose additional adverse ecological effects. In general, we noted that reforestation CSR campaigns across LMICs were deceitful as they predominantly focused on the number of trees given or planted (Chapman 1994) and rarely took into account species particularities, survival rates, extra care needs and local environmental health.

Perhaps due to these challenges, the tobacco industry's communications changed in the last half of the 1990s. As Geist et al. pointed out, reports issued by the International Tobacco Growers Association (ITGA, an organization funded by the industry) from that period stated that deforestation was not considered "a significant negative externality" and that the establishment of new energy-efficient and renewable sources of wood would stabilize the crop's impact on deforestation (1999, 19). Geist argues that this statement has to be challenged since it is based on a claimed change in the economics of fuel choice, which is put forward without sufficient data. The estimations produced in his study refuted ITGA's claims by showing that, "Deforestation related to tobacco constitutes an issue of global relevance which could be found on all continents, on average contributing nearly five percent to overall deforestation in the respective growing countries of the developing world (1999, 27)." Moreover, a more recent study on ecosystem impacts of tobacco farming (Geist et al. 2009, 1074) states that the industry cannot substantiate its claim that there has been "a continuous reduction in wood used for tobacco curing." The authors found that the 2009 rate of wood consumption in the Tabora district of Tanzania was no different from the rate observed in the same region 30 years before.

Knowledge gaps

The studies mentioned above sharply contradict the tobacco industry's discourse on deforestation and soil and water degradation. The Geist et al. (2009) study in particular begins to establish conditions for a standardized international comparison of the impacts of tobacco growing on ecosystems. However, so far no systematic process has been put in place to monitor impact and assess these global, cumulative effects. For example, the Specific Fuel Consumption measure (to measure fuel wood needs for curing) is still not widely used in research methodologies, which renders comparisons across studies and compilation of data difficult.

At a national level, the series of studies conducted in Tanzania effectively showed that fragile environments are prone to serious local impacts caused by tobacco production and accelerated deforestation. However, research that assesses the threats of tobacco growing on fragile environments remains scant in other tobacco-growing countries. Research results from Honduras (Loker 2005) and Bangladesh (Akhter et al., this volume) demonstrate a need for further research on the tobacco industry practices of resource mining in highly productive and fertile areas, and their associated ecosystem impacts. Finally, the industry discourse, through its corporate social responsibility campaigns and activities, needs to be thoroughly assessed to determine what actions are taken, or not taken, by the industry to resolve exploitative and damaging practices in tobacco-farming communities.

Conclusion

Transnational tobacco corporations have, in the past several decades, shifted their leaf production from HICs to LMICs, which has given them access to cheap labor and lowered production costs. Their business model is primarily based on the vertical integration system, often referred to as contract farming, which gives them direct access to a farmer base that produces tobacco leaf under the conditions that they set. Through this system, they are able to ensure both leaf quality and low prices by imposing their production requirements (the frequent use of agrochemical inputs) and grading scale.

In the last decade, the development of the WHO FCTC has threatened the tobacco industry's profits more than ever. To counter its ratification and implementation across the developing world, the industry has been very active in lobbying governments and intervening in policy making to block or water-down regulations that would threaten their market base and expansion in these countries. They insistently claim loud and clear (with the help of their front groups) that the implementation of FCTC policies is harmful to LMIC farmers, who, according to them, benefit highly from tobacco production and lack meaningful economic alternatives. At the same time, they have downplayed the negative impacts of tobacco farming in communities and have shifted attention away through inappropriate or weakly developed corporate social responsibility campaigns.

The literature review in this chapter and detailed case studies elsewhere in this volume show that most smallholder tobacco farmers are not benefiting economically from the crop. When a comprehensive calculation of production cost is done, including family labor and expensive agricultural inputs (generally advanced by tobacco companies), researchers find farmers' net profits to be significantly reduced. Despite the tobacco companies' provision of lump-sum payments at the end of the season (which sometimes creates the illusion of a high cash income) farmers have often been found to carry debts with the companies year after year. Power dynamics inherent to vertical integration helps to explain why farmers do not have the bargaining power to negotiate for higher prices and reduce debts. Other negative socioeconomic impacts include child labor, missed education opportunities and intra-household inequities affecting women in particular.

Research has also shown that tobacco-farming communities are faced with numerous occupational health hazards specific to tobacco production, namely exposure to molds and dust present in dried leaves, heavy metal particles absorbed by the plant, fuel wood smoke during curing and skin absorption of nicotine (green tobacco sickness, GTS). GTS affects a significant proportion of farmers and leads to many recurrent short-term symptoms that cause a great deal of discomfort and potentially unknown long-term and chronic impacts on health. Tobacco farmers also face a host of respiratory and other health problems caused by exposure to high levels of agrochemicals. While this exposure can be mitigated by using protective equipment and clothing, it has been found to be impractical in many settings and most farmers in LMICs do not have the capacity to purchase and manage the equipment.

Food insecurity is exacerbated by tobacco farming, especially among smallholders. This affects not only households and communities but also regions through the displacement of food crops on scarce arable lands. In many countries, the devotion of land to tobacco at a regional scale poses a food security risk, leaving both tobacco-growing and non-tobacco-growing communities vulnerable to unstable and uncertain markets for food. Food insecurity is also deepened by tobacco-farming practices that mine soil nutrients, pollute waterways and degrade forests and farmlands more severely than other commercial crops.

Recent research on alternative livelihoods to tobacco farming funded by the International Development Research Centre (and presented in this volume) has shown that smallholder farmers are receptive to shifting out of tobacco production when the conditions allow for it. In order to scale up these initiatives, however, results show that government policies and programs are needed to improve market structure, public extension services and subsidies and access to credit and loans for alternative crops. Tobacco farmers, just as all smallholder farmers in LMICs, need policy reforms that put agricultural development at the center of their public services. Recognizing that such policy reforms can take time, and that farming tobacco undermines the human and environmental health and economic growth of farming communities, the development of government programs designed specifically for the transition of tobacco farmers to alternative livelihoods are justified in the immediate term. Because of the influence that tobacco companies exercise in policy environments, what can accelerate tobacco control in the short term are both country case studies and global analyses of the industry's practices and strategies to undermine policy implementation. Such studies will contribute to further demystify the industry's claims about its corporate responsibility and the economic value of tobacco farming in LMICs and highlight its real agenda.

References

Abdallah, J. M. and G. G. Monela. 2007. "Overview of Miombo Woodlands in Tanzania." Working Papers of the Finnish Forest Research Institute 50: 9–23.

Abdallah, J. M., B. Mbilinyi, Y. N. Ngaga and A. Ok'ting'ati. 2007. "Impact of Flue-Cured Virginia on Miombo Woodland: A Case of Small-Scale Flue-Cured Virginia Production in Iringa Region, Tanzania." *Discovery and Innovation* 19 Supplement (1–2): 92–106.

Akhter et al., this volume.

Akhter, F. 2010. "Shifting out of Tobacco: Farmers' Initiatives to Grow Food Crops." UBINIG Research Findings, PowerPoint Presentation.

Akhter, F., F. Mazhar, M. A. Sobhan, P. Baral, S. Das Shimu and Z. Alam Khan. 2008. "From Tobacco to Food Production: Assessing Constraints and Transition Strategies in Bangladesh." Final Technical report submitted to the Research for International Tobacco Control (RITC) program of the International Development Research Centre (IDRC), Canada, 20 pp.

Almeida, G. E. G. 2008. "The Biopolitic of the Human Rights: A Reflection from Tobacco Contract System." [A biopolítica dos direitos humanos: uma reflexão a partir da sistema de integração rural da fumicultura.] Master's thesis, University of Brasília, Law Institute, 201 pp.

Almeida, this volume.

Amigó, M. F. 2010. "Small Bodies, Large Contribution: Children's Work in the Tobacco Plantations of Lombok, Indonesia." *The Asia Pacific Journal of Anthropology* 11(1): 34–51.

Arcury, T. A. and S. A. Quandt. 2006. "Health and Social Impacts of Tobacco Production." *Journal of Agromedicine* 11(3–4): 71–81.

Arcury, T. A., S. A. Quandt and S. Simmons. 2003. "Farmer Health Beliefs about an Occupational Illness that Affects Farmworkers: The Case of Green Tobacco Sickness." *Journal of Agricultural Safety and Health* 9(1): 33–45.

Babalola, A. 1993. "Capitalist Development in Agriculture: The Case of Commercial Tobacco Farming in the Oyo-North Division, Oyo State, Nigeria." *African Economic History* (21): 37–49.

Babalola, S. O. and C. Dennis. 1988. "Returns to Women's Labor in Cash Crop Production: Tobacco in Igboho, Oyo State, Nigeria." In *Agriculture, Women and the Land: The African Experience*, edited by J. Davison, 79–89. Boulder, Colorado: Westview Press.

Bortoluzzi, E. C., D. D. S. Rheinheimer, C. S. Gonçalves, J. B. Pellegrini, R. Zanella and A. C. C. Copetti. 2006. "Contamination of Surface Water by Pesticides as a Function of Soil Use in the Agudo Watershed, RS." [Contaminação de águas superficiais por agrotóxicos em função do uso do solo numa microbacia hidrográfica de Agudo, RS] *Revista Brasileira De Engenharia Agricola e Ambiental* 10(4): 881–87.

Bunnak, H. E. P., M. Kong and D. Yel. 2009. "Study on Tobacco Farming in Cambodia." Southeast Asia Tobacco Control Alliance, 54 pp.

Chapman, S. 1994. "Tobacco and Deforestation in the Developing World." *Tobacco Control* 3(3): 191 pp.

Cornwall, J. E., M. L. Ford, T. S. Liyanage and D. W. K. Daw. 1995. "Risk Assessment and Health Effects of Pesticides Used in Tobacco Farming in Malaysia." *Health Policy Plan* 10(4): 431–37.

Damalas, C. A., E. B. Georgiou and M. G. Theodorou. 2006a. "Pesticide Use and Safety Practices among Greek Tobacco Farmers: A Survey." *International Journal of Environmental Health Research* 16(5): 339–48.

Damalas, C. A., M. G. Theodorou and E. G. Georgiou. 2006b. "Attitudes towards Pesticide Labeling among Greek Tobacco Farmers." *International Journal of Pest Management* 52(4): 269–74.

Dowla, H. A., M. Panemangalore and M. E. Byers. 1996. "Comparative Inhibition of Enzymes of Human Erythrocytes and Plasma in Vitro by Agricultural Chemicals." *Archives of Environmental Contamination and Toxicology* 31(1): 107–14.

Espino, R. R. C., D. L. Evangelista and E. U. Dorotheo. 2009. "Survey of the Tobacco Growing Areas in the Philippines." South-East Asia Tobacco Control Alliance, 81 pp.

Fraser, A. I. 1986. "The Use of Wood by the Tobacco Industry and the Ecological Implications." Edinburgh: International Forest Science Consultancy, 20 pp.

Geist, H. J. 1999. "Global Assessment of Deforestation Related to Tobacco Farming." *Tobacco Control* 8: 18–28.

Geist, H. J., K. Chang, V. Etges and J. M. Abdallah. 2009. "Tobacco Growers at the Crossroads: Towards a Comparison of Diversification and Ecosystem Impacts." *Land Use Policy* 26(4): 1066–79.

Ghosh, S. K., J. R. Parikh and V. N. Gokani. 1979. "Studies on Occupational Health Problems During Agricultural Operation of Indian Tobacco Workers. A Preliminary Survey Report." *Journal of Occupational Medicine* 21(1): 45–47.

Gonçalves, C. S., D. dos S. Rheinheimer, J. B. R. Pellegrini and S. L. Kist. 2005. "Qualidade da água numa microbacia hidrográfica de cabeceira situada em região produtora de fumo." *Revista Brasileira de Engenharia Agrícola e Ambiental* 9(3): 391–99.

Griza, F. T., K. S. Karen Saldanha Ortiz, D. Geremias and F. V. Thiesen. 2008. "Avaliação da contaminação por organofosforados em águas superficiais no município de Rondinha- Rio Grande do Sul." *Química Nova* 31(7): 1631–35.

Guedes de Lima, R. G. 2007. "Development and Labor Relations in Southern Brazilian Tobacco culture." [Desenvolvimento e Relações De Trabalho Na Fumicultura Sul-Brasileira] *Sociologias* (18): 190–225.

Hamade, this volume.

Heald, S. 1991. "Tobacco, Time, and the Household Economy in Two Kenyan Societies: The Teso and the Kuria." *Comparative Studies in Society and History* 33(1): 130–57.

Hoang Van, M., G. Kim Bao, V. Vu Thi, T. Le Quynh, T. Nguyen Thu, T. Ngo Tri, T. Nguyen Hoang, A. Hoang Ngoc et al. 2010. "Health Problems, Health Costs and Health Beliefs Related to Tobacco Cultivation and Processing among Tobacco Farmers in Rural Vietnam." Final progress report, submitted to the Research for International Tobacco Control (RITC) program of the International Development Research Centre (IDRC), Canada, 42 pp.

Hu, T., Z. Mao, H. Jiang, M. Tao and A. Yurekli. 2007. "The Role of Government in Tobacco Leaf Production in China: National and Local Interventions." *International Journal of Public Policy* 2 (3/4): 235–48.

ICRISAT (International Crops Research Institute for the Semi-Arid tropics), Malawi. 2009. "Assessing Tobacco Handling and Storage by Smallholder Farmers in Malawi: A Case Study of Ukwe EPA (Lilongwe), Chiosya EPA (Mchinji) and Chipala EPA (Kasungu)." Report submitted to the Research for International Tobacco Control (RITC) program of the International Development Research Centre (IDRC), Canada, 17 pp.

ILRF (International Labor Rights Forum). 2008. "Request for Information for the Development and Maintenance of the List of Goods from Countries Produced by Child Labor or Forced Labor." Federal Register Notice Vol. 72(247): 73374.

Kibwage, J. K., G. W. Netondo, A. J. Odondo and G. M. Momanyi. 2008. "Diversification of Household Livelihood Strategies for Tobacco Small-Holder Farmers: A Case Study of Introducing Bamboo in South Nyanza Region, Kenya." Third Interim Technical Report submitted to the Research for International Tobacco Control (RITC) program of the International Development Research Centre (IDRC), Canada, 35 pp.

Kibwage, J. K., G. W. Netondo, A. J. Odondo, G. M. Momanyi, A. H. Awadh and P. O. Magati. 2009. "Diversification of Household Livelihood Strategies for Tobacco Small-holder Farmers: A Case Study of Introducing Bamboo in South Nyanza Region, Kenya." Final Technical Report submitted to the Research for International Tobacco Control (RITC) program of the International Development Research Centre (IDRC), Canada, 25 pp.

Kibwage et al., this volume.

Kimura, K., K. Yokoyama, H. Sato, R. B. Nordin, L. Naing, S. Kimura et al. 2005. "Effects of Pesticides on the Peripheral and Central Nervous System in Tobacco Farmers in Malaysia: Studies on Peripheral Nerve Conduction, Brain-Evoked Potentials and Computerized Posturography." *Industrial Health* 43(2): 285–94.

Kirk, C. 1987. "Contracting Out: Plantations, Smallholders and Transnational Enterprise (Sri Lanka)." *Institute of Development Studies Bulletin* 18(2): 45–51.

Kweyuh, P. H. M. 1994. "Tobacco Expansion in Kenya: The Socio-ecological Losses." *Tobacco Control* 3(3): 248.

Lecours, N., G. E. G. Almeida, J. J. Abdallah and T. Novotny. 2012. "Environmental Health Impacts of Tobacco Farming: A Review of the Literature." *Tobacco Control* 21(2): 191–96.

Loker, W. M. 2005. "The Rise and Fall of Flue-Cured Tobacco in the Copan Valley and its Environmental and Social Consequences." *Human Ecology* 33(3): 299–327.

Lonsway, J. A., M. E. Byers, H. A. Dowla, M. Panemangalore and G. F. Antonious. 1997. "Dermal and Respiratory Exposure of Mixers/Sprayers to Acephate, Methamidophos, and Endosulfan during Tobacco Production." *Bulletin of Environmental Contamination and Toxicology* 59(2): 179–86.

Madeley, J. 1993. "The Environmental Impact of Tobacco Production in Developing Countries." *Proceedings of the Fifth World Conference on Smoking and Health*, Winnipeg, Canada 2: 287–90.

Mangora, M. M. 2005. "Ecological Impact of Tobacco Farming in Miombo Woodlands of Urambo District, Tanzania." *African Journal of Ecology* 43(4): 385–91.

McBride, J. S., D. G. Altman, M. Klein and W. White. 1998. "Green Tobacco Sickness." *Tobacco Control* 7(3): 294–98.

McKnight, R. H. and H. A. Spiller. 2005. "Green Tobacco Sickness in Children and Adolescents." *Public Health Reports* 120(6): 602–606.

Merten, G. H. and J. P. Minella, 2006. "Impact on Sediment Yield Due to the Intensification of Tobacco Production in a Catchment in Southern Brazil." *Ciência Rural* 36: 669–72.

Muller, M. 1976. *Tobacco in the Third World: Tomorrow's Epidemic?* London: War on Want.

Muwanga-Bayego, H. 1994. "Tobacco Growing in Uganda: The Environment and Women Pay the Price." *Tobacco Control* 3: 255–56.

Nakamura, T. 1984. "Tobacco Dermatitis in Japanese Harvesters." *Contact Dermatitis* 10: 310–18.

Nefedov, V. B., L. A. Popova and Z. Z. Zhalolov. 1991. "Lung Function in Tobacco Growers Suffering from Exogenous Allergic Alveolitis (in Russian)." *Ter Arkh* 63(3): 124–26.

Nguyen Thanh, H., M. Hoang Van, G. Kim Bao and L. Nguyen Tuan. 2009. "Impact of Tobacco Growing on the Livelihood and Health of Tobacco Farmers and the Environment: A Preliminary Study in Vietnam." Southeast Asia Tobacco Control Alliance, 72 pp.

Ohayo-Mitoko, G. J. A., D. J. J. Heederick, H. Kromhout, B. E. O. Omondi and J. S. M. Beleij. 1997. "Acetylcholinesterase Inhibition as an Indicator of Organophosphate and Carbamate Poisining in Kenyan Agricultural Workers." *International Journal of Occupational and Environmental Health* 3(3): 210–20.

Olade, R. and K. D. Lessnau. 2006. "Tobacco Worker's Lung." WebMD, emedicine. http://www.emedicine.com/med/topic2282.htm (accessed 12 December 2013).

Osim, E. E., C. T. Musabayane and J. Mufunda. 1998. "Lung Function of Zimbabwean Farm Workers Exposed to Flue Curing and Stacking of Tobacco Leaves." *South African Medical Journal* 88(9): 1127–31.

Otañez and Graen, this volume.

Otañez, M. G., M. E. Muggli, R. D. Hurt, and S. A. Glantz. 2006. "Eliminating Child Labor in Malawi: A British American Tobacco Corporate Responsibility Project to Sidestep Tobacco Labor Exploitation." *Tobacco Control* 15: 224–30.

Pain, A., I. Hancock, S. Eden-Green and B. Everett. 2012. "Research and Evidence Collection on Issues Related to Articles 17 and 18 of the Framework Convention on Tobacco Control." Report by DD International for British American Tobacco, 205 pp. Online: http://ddinternational.org.uk/viewProject.php?project=21 (accessed 7 December 2012).

Panemangalore, M., H. A. Dowla and M. E. Byers. 1999. "Occupational Exposure to Agricultural Chemicals: Effect on the Activities of Some Enzymes in the Blood of Farm Workers." *International Archives of Occupational and Environmental Health* 72(2): 84–88.

Plan Malawi. 2009. "Hard Work, Long Hours and Little Pay: Research with Children Working on Tobacco Farms in Malawi." Plan International, UK. Online: http://plan-international.org/files/Africa/RESA/malawi/Plan%20Malawi%20 child%20labour%20and%20tobacco%202009.pdf/view?searchterm=tobacco (accessed 12 June 2013).

Rao, P., S. A. Quandt and T. A. Arcury. 2002. "Hispanic Farmworker Interpretations of Green Tobacco Sickness." *Journal of Rural Health* 18(4): 503–11.

Riquinho, D. L. and E. A. Hennington. 2012. "Health, Environment and Working Conditions in Tobacco Cultivation: A Review of the Literature." *Ciência & Saúde Coletiva* 17(6): 1587–1600.

Salvi, R. M. D. R. Lara, E. S. Ghisolfi, L. V. Portela, R. D. Dias and D. O. Souza. 2003. "Neuropsychiatric Evaluation on Subjects Chronically Exposed to Organophosphate Pesticides." *Toxicology Science* 72: 267–71.

Samrech, P. 2008. "Survey of Tobacco Farming in Cambodia." Southeast Asia Tobacco Control Alliance, 65 pp.

Sauer, J. and J. M. Abdallah. 2007. "Forest Diversity, Tobacco Production and Resource Management in Tanzania." *Forest Policy and Economics* 9(5): 421–39.

Schmitt, N. M., J. Schmitt, D. J. Kouimintzis and W. Kirch. 2007. "Health Risks in Tobacco Farm Workers – A Review of the Literature." *Journal of Public Health* 15(4): 255–64.

Sejjaaka, S. 2004. "From Seed to Leaf: British American Tobacco and Supplier Relations in Uganda." In *International Businesses and the Challenges of Poverty in the Developing World*, edited by Frederick Bird and Stewart Herman, 111–23. New York: Palgrave Macmillan.

Southeast Asia Tobacco Control Alliance (SEATCA). 2008. "The Tobacco Trap: Cycle of Poverty in ASEAN Countries" (video). Southeast Asia Tobacco Control Alliance.

Szarmachz, H. and H. Poniecka. 1973. "Contact Allergy in Agriculture" (in Polish). *Przegl Dermatol* 60(4): 479–84.

Tobacco-Free Kids. 2001. "Golden Leaf, Barren Harvest: The Costs of Tobacco Farming." Campaign for Tobacco-Free Kids, 42 pp.

Tobin, R. J. and W. I. Knausenberger. 1998. "Dilemmas of Development: Burley Tobacco, the Environment and Economic Growth in Malawi." *Journal of Southern African Studies* 24(2): 405–24.

Trapé-Cardoso, M., A. Bracker, D. Dauser, C. Oncken, L. V. Barrera, B. Gould et al. 2005. "Cotinine Levels and Green Tobacco Sickness among Shade Tobacco Workers." *Journal of Agromedicine* 10(2): 27–37.

Vargas, M. A. and R. R. Campos. 2005. "Crop Substitution and Diversification Strategies: Empirical Evidence from Selected Brazilian Municipalities." The World Bank: HNP discussion paper, Economics of Tobacco Control Paper no. 28, 33 pp.

Vargas, M. A., M. Lootty, R. M. Alievi, B. Ferreira de Oliveira, B. Guimarães and R. B. A. Vargas. 2009. "The Impact of Tobacco Farming on Local Development Strategies in Brazil: Empirical Evidences of Crop Substitution and Diversification in the Rio Pardo Valley Region." Final progress report, submitted to the Research for International Tobacco Control (RITC) program of the International Development Research Centre (IDRC), Canada, 33 pp.

Waluye, J. 1994. "Environmental Impact of Tobacco Growing in Tabora/Urambo, Tanzania." *British Medical Journal* 3(3): 252.

WHO (World Health Organization). 1980. "Save the Rain Forests." *IUCN Bulletin*, 11(5).

Yanda, P. Z. 2010. "Impact of Small Scale Tobacco Growing on the Spatial and Temporal Distribution of Miombo Woodlands in Western Tanzania." *Journal of Ecology and the Natural Environment* 2(1): 10–16.

Section Three

ECONOMICALLY SUSTAINABLE ALTERNATIVES TO TOBACCO

Tobacco Industry Myth: There are currently no economically sustainable alternatives to tobacco farming for small-scale farmers, particularly in low- and middle-income countries.

Research Findings:

- Where support for smallholder agriculture was nonexistent or had been reduced (particularly as a result of structural adjustment programs), many farmers felt they had little choice but to turn to tobacco where the industry provides the missing market infrastructure and extension services needed.
- Despite this challenge, as the case studies here show, many other crops, crop combinations, farming systems and livelihood strategies offer better opportunities for farmers.
- Although the current generation of tobacco farmers will not be affected by tobacco-control measures, given the harsh nature of the work, it will be important for governments to help farmers transition to alternative crops.
- The transition will require a national vision for sustainable rural development. Such a vision should include support for market infrastructure and extension services for alternative crops that were previously missing. Furthermore, access to public financing for tobacco-farming transitions is justified, and can be financed in part through domestic taxes on the consumption of tobacco and, where it exists, the removal of public funding for tobacco cultivation.
- The active participation of farmers and other stakeholders in the development of new options is key to success.

Chapter 5

BREAKING THE DEPENDENCY ON TOBACCO PRODUCTION: TRANSITION STRATEGIES FOR BANGLADESH

Farida Akhter, Daniel Buckles and Rafiqul Haque Tito[1]

Introduction

Tobacco farmers in Bangladesh are, in their own way, as dependent on tobacco as smokers of the final product. Debt to the tobacco companies, and the seductive appeal of facilities they offer, bind tobacco farmers to an industrial monocrop that depletes soils, denudes forested hillsides and compromises the health of field workers, and of the women and children curing the leaves (see Lecours, this volume). Many tobacco farmers, especially older ones who have seen the impacts of tobacco growing on their families and on their lands, are desperate to shift to other crops, but feel they cannot. Local and regional markets have withered in tobacco-growing regions, locally adapted seeds for food crops are not readily available and soils are so degraded by years of tobacco cultivation that to grow any crop at all seems impossible without using massive amounts of fertilizers and pesticides. Breaking the dependency on tobacco production is not easy, and many farmers that consider it find themselves going back to the tobacco companies year after year.

The Government of Bangladesh, as a party to the World Health Organization–sponsored Framework Convention on Tobacco Control (FCTC), has indicated that it intends to help farmers shift out of tobacco production. Doing so without excessive costs to governments or hardship

1   The authors would like to gratefully acknowledge the research support of UBINIG staff, farmer leaders of the Nayakrishi Andolon and the IDRC. The authors remain responsible for the arguments and any errors.

for farmers remains a challenge, however. The Smoking and Tobacco Products Usage (Control) Bill passed by the Bangladesh government in 2005 included provisions to support alternative crops (Article 12). A proposed 2013 amendment to the bill will focus on removing all government incentives to tobacco production in food-producing areas.[2] The amendment emerged in response to rapid increases in the land area under tobacco cultivation in Bangladesh, concerns about the diversion of land and agricultural inputs from food production to tobacco and mounting evidence of the impacts of tobacco growing on farmers and sensitive forest resources. Little detailed research has been done, however, on the obstacles to change that Bangladeshi tobacco farmers face or the practical strategies governments can use to support or revive food production in tobacco-growing areas.

A participatory action research initiative launched in 2006 by UBINIG, a Bangladeshi policy and action research organization, and Carleton University (Ottawa, Canada), seeks to address this gap. This chapter presents some of its findings, with a focus on the factors and actors enabling a transition out of tobacco production.[3] The chapter starts by examining the causes and conditions that led to expansion of the Bangladesh tobacco industry in the first place. This is followed by a detailed description of tobacco farmers in two important tobacco-growing areas: Kushtia in central Bangladesh and Cox's Bazar and Bandarban along the Matamuhuri River in southeastern Bangladesh (Chittagong). Farmers in these two areas contributed to a collaborative assessment of the reasons why they continue to grow tobacco and the constraints they face. The analysis of these constraints includes a discussion of the tactics of British American Tobacco (BAT) and other national tobacco companies in Bangladesh that create obstacles to change. A separate section examines the process and results of collaborative research with farmers aimed at developing and assessing regionally adapted transition strategies, and efforts to implement the strategies at a larger scale.

The Evolution of Tobacco Production in Bangladesh

The emergence and expansion of tobacco production in Bangladesh is not a farmer-led phenomenon. Rather, it is driven by international and national tobacco companies' desire to secure a steady supply of tobacco leaf for the manufacture of cigarettes and related tobacco products. As discussed in this

---

2   Media reporting on the proposed amendment was significant in March 2013, with 37 articles published in various weekly and daily media outlets (Karim 2013).

3   A comprehensive collection of findings from the research has also been published in Bangla (Akhter et al. 2012).

section, the evolution of tobacco production in Bangladesh has been driven by broader political and economic conditions that created a vacuum in the agricultural sector into which tobacco farming was inserted. It has also followed a pattern of shifting the production of Flue-Cured Virginia (FCV) varieties of tobacco from one part of the country to another as soil and forest resources are mined and depleted.

Tobacco as an industrial crop has its roots in the mid-1960s in East Pakistan when the BAT launched experiments with the Flue-Cured Virginia variety of tobacco suitable for cigarette manufacture (Maniruzzaman et al. 2011). In the years immediately after Independence in 1971, the BAT began to produce FCV for the national and international markets, making use of the then-fertile Teesta River silt soils of Rangpur in northern Bangladesh. The plant thrived on these sandy, well-aerated and well-drained soils, especially during the cooler winter season. Over time, national cigarette companies emerged as players in specific markets, including the purchase of burley tobacco (air-cured or sun-dried tobacco for cigarettes sold in the United States) and the dark air or shade-cured tobacco (Jati and Motihari varieties) used in bidi (hand-rolled) cigarette manufacture, hookah paste, chewing tobacco and other tobacco products for the national market. At present, four companies – BAT, Abul Khair Leaf Tobacco, Dhaka Tobacco Company and Nasir Tobacco Industries Ltd. – account for virtually all of the export production and a large share of the national market (Akhter et al. 2012). Some five percent of BAT's global tobacco production comes from Bangladesh (Pain et al. 2012).

Two national policies were particularly instrumental in creating favorable conditions for emergence of the tobacco industry in Bangladesh: the promotion of foreign direct investment (FDI) and crop diversification programs. From the 1960s to the present day, conditions placed on loans from the Bretton Woods Institutions and from many bilateral aid agencies fostered FDI and promoted foreign companies' access to certain segments of the agriculture sector in Bangladesh (Sobhan 1995). The intention was to boost economic growth by attracting multinational private corporate investment and increasing cash export earnings, with the quantity of the investment as a primary consideration. Little or no consideration was given to the nature of the investment or its potential effects on other sectors of the economy, the environment and matters of national food security. In this policy environment, investments by seed companies and by tobacco companies were seen as positive to Bangladesh's development. These two corporate activities (seed and tobacco) have accounted for a significant proportion of FDI in the agricultural sector since Independence.

FDI as a structural adjustment policy dovetailed particularly well during the 1980s and 1990s with crop diversification programs funded by bilateral donors such as the Canadian International Development Agency (now the Department of Foreign Affairs, Trade and Development), the United States' Agency for International Development (AID) and various European agencies. These actors exercised considerable influence in the Bangladesh economy, and particularly in agriculture. While ostensibly aimed at reducing the dominance of rice in Bangladesh agriculture, crop diversification programs of the time also created new opportunities for the tobacco industry by including tobacco among the range of cash crops eligible for subsidies of various kinds (Hoque 2001). This directed scarce public resources into providing tobacco producers with subsidized access to urea fertilizers, irrigation water, technical assistance and basic agricultural research – services that could have been allocated to essential food crops.

The inclusion of tobacco in crop diversification programs helped to create a positive image and environment for tobacco company operations. It allowed them to consciously position their production in plain sight among the common cash crops of Bangladesh, even as smoking and consumption of tobacco products became recognized on the international stage as severe health threats. This legitimacy could not be achieved so easily today. While still a legal crop in Bangladesh, tobacco production has little in common with jute, oil seeds, spices and vegetables – cash crops that every Bangladeshi welcomes and recognizes as making positive contributions to society.

The emergence of tobacco production on the national scene in Bangladesh is reflected in data on production trends (Figure 5.1). Tobacco production increased gradually from the 1960s until the early 1980s, both in terms of area cultivated and metric tons of tobacco leaf grown. This represents a period of expansion without technological innovation. During the 1990s total production measured in tonnage increased even as the total land area dedicated to tobacco remained relatively stable. The productivity gains, reflected in yield per hectare (ha), can be attributed to improved tobacco seed, new cultivation methods, a higher degree of fertilizer use and more chemical pest control. A severe viral attack in 1998 sent production to its lowest point since the early 1960s, but this was quickly turned around with an anti-viral product provided by international agrochemical companies (UBINIG 2010). Production has climbed steadily ever since, and very sharply in the last few years. As discussed later in this chapter, the rapid increase in tobacco production since 2009 has come at the direct expense of food production and watershed protection in some of Bangladesh's most productive and vulnerable agricultural settings.

Most of the recent growth in national tobacco production is due to an increase in FCV, which saw its share of total tonnage climb from 29 percent in 1995 to 64 percent in 2009. While tobacco for bidi production remained stable in absolute terms, it has declined dramatically as a share of total tobacco production. This reflects an industry shift towards FCV production for export. The timing for this

Figure 5.1. Tobacco leaf production in hectares (ha), metric tons (MT) and hectograms (hg) per hectare, 1961–2011, Bangladesh

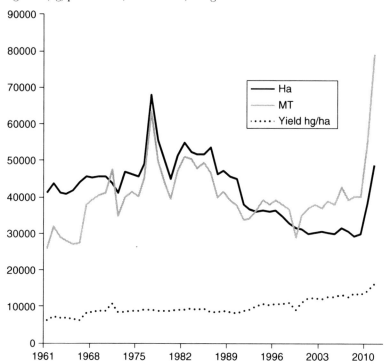

Source: FAO (2012). FAOSTAT Production dataset. Online: http://faostat3.fao.org/home/index.html (accessed 20 August 2012).

(early 1990s to the present) follows the international pattern of shifting production sites from centers of tobacco supply in the high-income countries to farms in low- and middle-income countries. Over the last 20 years, international tobacco companies have moved out of supply environments where they have faced increased regulation for tobacco farming to settings where regulation is weak and governments are more easily influenced (FAO 2003; Buckles et al., this volume).

While aggregated national data sheds light on external conditions and national policy factors influencing the evolution of tobacco production in Bangladesh, it masks another key driver – the mining of soil and forest resources by the tobacco industry and periodic shifts to new extraction sites. Data on FCV tobacco production over time and estimated productivity indicators from three important tobacco-growing districts in Bangladesh – Rangpur, Kushtia and Bandarban – reveals the pattern. Between 1995 and 2011, the area in hectares dedicated to FCV in Rangpur remained steady while it increased rapidly in Kushtia (Figure 5.2). During this same period,

Figure 5.2. Hectares of Flue-Cured Virginia (FCV) tobacco in Rangpur, Kushtia and Bandarban, 1995–2011

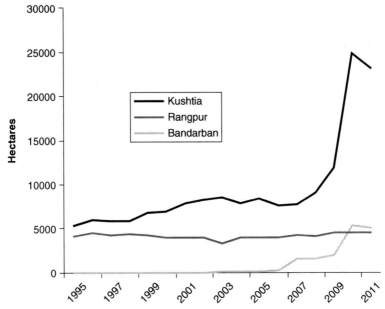

Source: Bangladesh Bureau of Statistics (2010).

yield per hectare increased by only nine percent in Rangpur compared to an increase of 57 percent nationally and an increase of 93 percent in Kushtia. These production and productivity trends reflect a shift in production for export-oriented FCV tobacco over time from Rangpur, where the tobacco industry started, to Kushtia. More recently, total production of FCV tobacco in Bandarban went from almost nothing in 1995 to more than 5,300 hectares in 2011, with yields per hectare well above the national average. Production of dark air varieties of tobacco (Motihari and Jati) also increased in Bandarban during this same period.

The shift in production sites can be explained by the declining quality of soil and forest resources needed by the tobacco industry. Once the center of tobacco production in Bangladesh, Rangpur has been virtually abandoned by international tobacco companies, leaving national companies to produce lower grade dark air and shade-cured tobacco suitable for national bidi markets.[1] Rangpur remains an important site for bidi production but no longer

---

4   Research on bidi-dependent livelihoods in Bangladesh (Roy et al. 2012, 314) shows that the industry deepens poverty among bidi workers and bidi users. The majority of bidi workers are women and children classified as unpaid assistants, and among "the 40%

competes with other areas of the country in the production of export-oriented FCV tobacco due to the generalized decline in soil fertility and fuel sources. According to a BAT-sponsored study (Pain et al. 2012, 204), "British American Tobacco Bangladesh has gradually reduced its operations in Rangpur."

Signs of decline in the conditions needed to produce FCV tobacco are also emerging in Kushtia. Farmers in Kushtia note that after continuous tobacco cultivation the soil becomes hard, dries up quickly or does not drain easily. The natural smell of the soil disappears and soil color changes. These observations are indicators of the loss of soil organic matter, changes in soil chemical properties and the loss of water-holding capacity (Akhter et al. 2012). The parasitic plant Orobanche spp., known in Bangladesh as *mula*, has also become a common noxious weed in the district's tobacco fields. It grows quickly on the roots of the tobacco plant and dramatically reduces its growth. It also attacks the roots of many other plant species including a wide range of vegetables, pulses and pasture legumes (Lins et al. 2005; Eizenberg et al. 2012). Control of Orobanche spp. is difficult, and its continuous spread limits the choice of rotational crops (Abu-Irmaileh and Labrada nd). Farmers in tobacco-growing areas infested by Orobanche spp. can be forced to abandon the land for many years.

In addition to these soil problems, the sources of fuel wood are decreasing, which means the costs of curing tobacco leaves are increasing steadily in Kushtia. This is because farmers' fields and community forest resources within the district can no longer supply the fuel needed to cure tobacco. Farmer-managed processing relies instead on imports of firewood from other districts. Interviews with farmers in Kushtia indicate that when the tobacco industry became active there in the mid-1970s, curing of tobacco leaves relied entirely on firewood from community forests. By 1985, about 85 percent of the fuel used to cure tobacco was firewood, and the remainder jute sticks. Ten years later, the use of firewood for curing had dropped to about 25 percent, and farmers were compensating for this by using a mix of jute sticks and rice straw. From 2000 to the present, firewood disappeared altogether from the FCV processing system, replaced entirely by straw (about 85 percent) and jute (15 percent). Much of the straw is imported from Jessore District, and the jute from Faridpur District, with a mix of both coming from Magura District. Very little fuel is taken from sources within Kushtia or from tobacco farmers' fields and community forests (Akhter et al. 2012). This fact prompted a BAT-commissioned study to conclude, "tobacco was not a cause of deforestation

---

of the Bangladeshi population living below the international poverty line of USD 1.25 per day." Virtually all bidi users are also poor, yet spend almost 10 percent of their daily income on tobacco.

Photograph 5.1. Woman in Kushtia feeding the fire of a tobacco kiln with rice straw while carrying her baby. Curing takes periods of 60–70 hours of continuous tending without sleep.

Photo credit: Abdul Zabbar.

in Kushtia" (Pain et al. 2012). The claim is only true, however, in the most myopic sense – trees and other sources of fuel in Kushtia were used up over a period of 20 years before the study, leaving tobacco farmers with no choice but to satisfy the need for fuel from external sources. Tobacco production today is not a cause of deforestation in the district because the tobacco-induced deforestation and mining of agricultural biomass in Kushtia occurred decades before and has now shifted to other districts.

The collapse of growing conditions in Rangpur and the gradual decline in Kushtia have made BAT and other tobacco exporters turn their attention to the richly forested and highly productive lands of southeastern Bangladesh. Bandarban District and the eastern edge of Cox's Bazar District in particular have experienced very rapid increases in the area dedicated to FCV tobacco production (Figure 5.2). Much of this growth is along the fertile flood plain of the Matamuhuri River, the only major watershed fully contained within the political boundaries of Bangladesh and consequently a water source of strategic long-term value (Baset 2011; Haque Tito 2010). Along a distance of more

Photograph 5.2. Tobacco grown on the shores of the Matamuhuri River, with denuded hills of the Chittagong Hill Tracts in the background

Photo credit: Abdul Zabbar.

than 80 kilometers, both banks of the river, renewed annually by nutrient-rich alluvial deposits, are now taken over by tobacco production. Chakaria in Cox's Bazar District and Ali Kadam and Lama *upazilas* (sub-districts or counties) in Bandarban District are dominated by tobacco production. This includes the many small islands and new surfaces (*char*) created each year through the process of accretion along riverbanks. The *upazilas* are also close to abundant sources of firewood in the Chittagong Hill Tracts. Field studies show that thousands of tons of firewood are brought down from the Chittagong Hill Tracts to the tobacco-growing sites every year, making use of the river for transportation (UBINIG 2009; Bala 2010). The river ecosystem is subject to heavy siltation provoked by the deforestation, flash flooding along the riverbank and pollution due to runoff from fertilizer and pesticide residues used in tobacco production.

Land lease prices reflect the many advantages of lands along the Matamuhuri river. In Ali Kadam and Lama *upazilas* in Bandarban District, the lease value per *bigha* along the banks of the Matamuhuri River is BDT 12,300 to BDT 16,600. Yields on these lands are 500 to 600 kg per *bigha* (UBINIG 2011). By contrast, in Kushtia the cost to lease agricultural lands for tobacco is less than half, that is, BDT 5,000 to BDT 7,000 per *bigha* (about 0.133 ha). In a normal year, farmers can expect between 400 and 450 kg of dried tobacco leaves from this area.

Until recently, the intrusion of the tobacco industry into southeastern Bangladesh passed largely unnoticed due to its relative isolation. The future scenario for the districts in the Chittagong Hill Tracts bordering the Matamuhuri River is already evident, however, in the damaged landscapes of Rangpur and Kushtia. Moreover, the mining of soil and forest resources that characterizes the evolution of tobacco production in Bangladesh is now taking advantage of at least three vulnerable *adivasi* (tribal) communities in this region. As tobacco production moves further and further up the Matamuhuri River it enters directly into the territories of the Marma, Muru and Chakma in the Chittagong Hill Tracts, tribal communities already exposed to exploitation, displacement and cultural assimilation. Tobacco brings new security risks to Bangladesh by undermining the viability and sustainability of food-producing communities in the Chittagong Hill Tracts. As we argue below, the extractive and destructive history of the tobacco industry in Bangladesh fully justifies restricting the expansion of tobacco production into new areas of the country.

Why Do Farmers Grow Tobacco?

Why do farmers in Bangladesh continue to grow tobacco, despite the many concerns they have about its impacts on their health, the environment and the land? To answer this question we must first ask, "Who are the tobacco farmers?"

It is difficult to get a precise estimate of the number of tobacco farmers in Bangladesh because the Bangladesh Bureau of Statistics (BBS) does not disaggregate data on the number of farming households growing a particular crop. Tobacco industry representatives in Bangladesh put the total number of tobacco farmers at around 100,000 (Rahman 2010), although it is not clear if this refers only to farmers contracted directly by tobacco companies or also includes tobacco farmers without company contracts.

Contract farming is a relatively recent trend in many agricultural sectors but it has a long history in the tobacco industry. Tobacco production in Bangladesh was vertically integrated through contractual arrangements with farmers from the start. Companies provide contract growers with loans they can use to prepare their land or lease land as needed. The company card associated with the contract also provides growers with access to credit for inputs (seed of a particular tobacco variety, selected fertilizers, fungicides, pesticides and sucker control chemicals, as well as other inputs such as polyethylene wrap for transplanting). The companies fix prices as a package, rather than charging farmers a unit price for each input. This obscures real costs and impedes competitive buying. The companies also buy cured tobacco leaf from the contracted farmers at fixed prices subject to grading at the time of delivery. The company contract specifies the quota amount of leaf agreed to and a delivery date subject to action by the company if not met. These arrangements resemble piecework found in a factory setting where all production decisions are made by factory management, not by the workers themselves. This system carries significant additional risks, however, since loans and credit purchases that farmers do not repay with the harvest are carried forward to the following year, with high levels of interest. In bad years, farmers under contract accumulate debt. This can lead over time to a cycle of debt bondage.

Photograph 5.3. Inside the tobacco-buying houses women do much of the work sorting and packing the tobacco leaves

Photo credit: Abdul Zabbar.

Contract tobacco growing often relies on and actively reinforces land lease markets in tobacco-growing regions. Land leasing from absentee or large landowners is common in Kushtia and Bandarban, accounting for as much as 50 and 70 percent, respectively, of the land where tobacco is grown. Landless households with a company card make use of the arrangement to lease land, as do smallholders. Arrangements involving land leasing and subcontracts with cardholders create the same web of obligations, including credit repayment and the company-provided production technology. High land prices for tobacco lands are a primary cause of indebtedness among tobacco farmers that lease land.

Leaseholders often shift from one plot to another when the land they are leasing becomes too poor to grow tobacco or becomes infested with persistent weeds such as *mula*. This practice reproduces at the micro level the industry pattern of shifting from one production site to another as field conditions deteriorate. Much like the tobacco companies, tobacco farmers with leased land have little stake in the health of the land beyond the period when their crop is growing.

The business of leasing land for tobacco cultivation has exacerbated problems associated with *khas* lands in Bangladesh. This type of land tenure refers to the relatively fertile *char* areas (islands and new riverbanks created through accretion of alluvial deposits) and lands seized by the government from individual properties that exceed the land reform limit of 14 hectares. In theory, *khas* lands, totaling some 1.3 million hectares in Bangladesh, are the property of the government and reserved for lease to the landless. In fact, they are often used illegally for the benefit of local elites (Barkat et al. 2000). This is particularly problematic in Bandarban where indigenous land tenure systems are already complex and local authorities easily influenced by elite groups. While this practice has been challenged recently (see below), *khas* lands remain an important avenue for accessing land for tobacco production in a number of areas.

The land tenure dynamics involved in the tobacco industry are further complicated by practical limits on the scale of tobacco production. Tobacco production units in Bangladesh tend to be less than one hectare and similar in size among both smaller and larger tobacco farms. This shows that tobacco cultivation is not simply an occupation of larger and better-off farmers, as the industry sometimes suggests (Pain et al. 2012). In a non-mechanized environment such as Bangladesh, tobacco must be managed at a small-scale, whether the farmer has extensive land or not. This practical limit on the scale of production at the farm level is due to the labor-intensive nature of tobacco farming and limitations on the storage of tobacco in the home, where leaves are dried and packed for sale. These tasks are done by hand, often from within the household labor pool or with the benefit of periodic hired labor. To grow more than a hectare of tobacco, and manage the labor and other resources such as fuel wood and storage needed to produce a final product for sale, is challenging for a single household. Consequently, tobacco farmers typically limit the size of their tobacco plots and, if they can, engage in tobacco trading and land leasing to realize other gains from the industry. This dynamic takes place in a national context where farm landholdings are already very small, only one hectare on average (Bangladesh Bureau of Statistics 2010).

The land dedicated to tobacco production serves no other purpose, a serious loss of potential from a food production perspective. The tobacco system is a long duration monocrop that interferes directly with virtually all other crops grown in the region, at the time of harvest, time of planting or both (Figure 5.3). Starting with the seedling stage of production and finishing with the harvesting and processing of the last ripened leaves, the tobacco season spans a full seven months, from early October to the end of April. This period overlaps sowing and harvesting for virtually all cereal crops, the sowing period for major jute varieties, sowing and harvesting for many pulses and spice crops and all winter vegetable crops including cash crops such as potatoes. As a result, land under tobacco makes poor use of the three distinct growing seasons in Bangladesh (the *rabi* season from January to April, the *kharif 1* season from May to August and the *kharif 2* season from August to December). The potential for triple cropping created by the country's climate and some land types is thus denied by tobacco farming, making the land use fundamentally inefficient. This has serious implications for food security in Bangladesh, as discussed below. The high lease value of land dedicated to tobacco also has a negative effect on the availability of land for food crops by driving up land prices beyond what food farmers can afford.

Figure 5.3. The crop calendar for tobacco and major food crops of Bangladesh

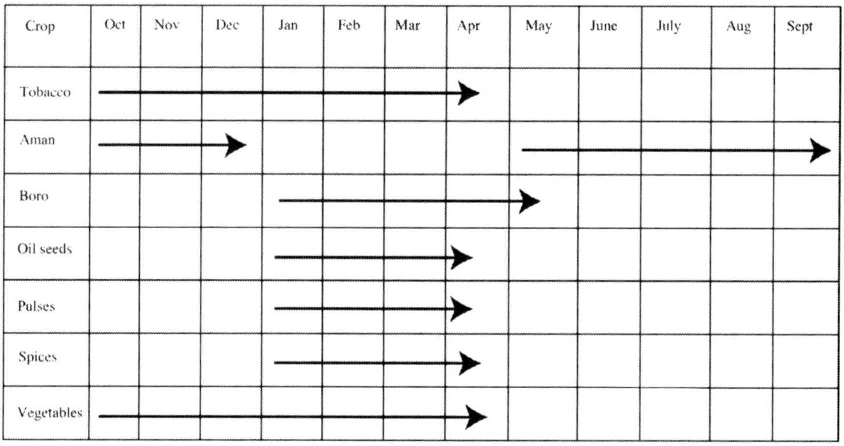

Crop	Oct	Nov	Dec	Jan	Feb	Mar	Apr	May	June	July	Aug	Sept
Tobacco	———	———	———	———	———	——→						
Aman	———	———	——→						———	———	———	——→
Boro				———	———	———	——→					
Oil seeds				———	———	——→						
Pulses				———	———	——→						
Spices				———	———	——→						
Vegetables	———	———	———	———	———	——→						

As described above, the interplay of contract farming, land leasing, plot size and cropping season is complex. It also has a paradoxical effect on the industry and the people involved. On the one hand, farmers are subject to strong processes of integration into the industry through formal contracts involving controlled markets, required inputs and household-based processing. On the other hand, the production system is fragmented into many tiny units cutting across and disturbing virtually every other cropping option available to farmers. This dichotomy creates a wide range of different tobacco farmer profiles, including younger, specialized tobacco farmers, older farmers with very small plots, tobacco traders with limited or no land of their own, landless leaseholder farmers and older, land-rich farmers who also lease out land for tobacco farming and engage in the tobacco trade. Other stakeholders include wage workers in tobacco fields and suppliers of firewood for kilns.

Given the complexity of factors and actors within communities where the tobacco industry is prevalent, district-wide data on landholdings and contract growing is difficult to collect. Nevertheless, a UBINIG survey of five *upazilas* in three districts where tobacco is grown confirms that many different kinds of households depend on tobacco farming, including large and smallholders (Table 5.1). More than half of the growers are landless tenant farmers. Roughly two-thirds of all tobacco growers are under contract with companies, while the remainder are bound to companies indirectly through purchase arrangements with contract holders and landholders. These different household profiles should be taken into account when assessing why farmers continue to grow tobacco and the development of policies to support a shift out of tobacco farming.

Table 5.1. The distribution of tobacco farmers in five counties (*upazilas*), by landholding and relationship to the tobacco companies

Upazila	District	Total number of tobacco growers	Larger landowners	Smaller landowners	Tenant farmers	% of tobacco growers under contract
Daulatpur	Kushtia	12,955	2,073	3,109	7,773	60
Mirpur	Kushtia	9,233	1,478	2,216	5,539	65
Lama	Bandarban	5,833	936	3,499	1,398	63
Ali Kadam	Bandarban	1,186	191	284	711	70
Chakaria	Cox's Bazar	3,073	493	737	1,843	65
Total		32,280	5,171	9,845	17,264	

Source: UBINIG field survey (2008).

Farmer perspectives on causal factors

Analysis by the authors of household profiles, land holdings and contract farming provided the research team with a basis for convening households with different characteristics to discuss the reasons why farmers continue to grow tobacco. The assessment used a participatory method called System Dynamics adapted from input–output reasoning used in the field of economics (Chevalier and Buckles 2013). Mixed groups reflecting various household profiles sat together to discuss their perspectives and develop an overall assessment of the factors making it difficult to shift to other crops. The group analysis converged around potential entry points for action – what farmers and the research team could do together to overcome constraints tying farmers to tobacco production. Results from Daulatpur, a prominent tobacco-growing *upazila* in Kushtia District, illustrate the process and farmer reasoning (Buckles 2008).

Villagers in Daulatpur identified six reasons why people in their village continue to grow tobacco. Here are the views they expressed:

- The price set by British American Tobacco (BAT) for the highest grade of cured tobacco is attractive. We hope we will get the highest price (even though our leaves and prices are often down-graded).
- BAT issues a contract to farmers they buy from, and will only buy from farmers with a contract. We can use the contract to buy cured tobacco leaf from farmers that do not have a contract, setting ourselves up as tobacco traders as well as producers.
- We receive a single payment for our entire crop. This is attractive because we can use the lump sum to repay debt, buy land, improve our houses or pay for other large expenditures such as weddings and other social events. We can also use the cash to buy tobacco from other farmers and thereby join the tobacco trade. Large lump sums of money paid are not available for other crops.
- There are currently few cash crops for us to consider or compare with tobacco. Markets for traditional cash crops (lentils, pulses and spices) have withered away over the years and no new cash crops have emerged.
- Most farmers in this area grow only tobacco. We feel peer pressure to farm this way because it is what all farmers do here.
- Tobacco cultivation makes use of family labor, especially women and children. By tending the fires to cure the tobacco leaves, women and children contribute directly to generating household cash income.

Men and women in the group discussed each of these factors and other factors they set aside as being less important. They did so with a focus on the relative weight of each factor and interactions between the factors at play. The key question posed for each factor was simple: to what extent does it contribute to other factors in the list? In other words, what is the causal weight of one factor on another? To answer the question each time it was asked, the group used a scale from one (indicating a very minor weight) to five (indicating a major causal weight). The group justified and negotiated ratings for each factor's causal interaction with other factors until a consensus emerged. The facilitators used a tree metaphor with roots and branches to support the discussion and help people keep track of which relationship they were talking about — is it A causing B, or B causing A? The numbers reflect the complex rationale that farmers have for continuing to grow tobacco (Table 5.2).

Interpretation of the results by participants focused on a Cartesian graph showing the location of each factor based on the intersection of the Cause Index and the Effect Index from the table (Figure 5.4). While the graph was an unfamiliar image to many of the participants, discussion brought out the meaning. Collectively, they saw that many of the reasons for growing tobacco are both causes and effects of each other (upper right quadrant of the figure), that is, root causes and their own ramifications at the same time. This includes the lump-sum payment, price for best grade tobacco, dominance of the crop and the narrow range of alternative cash crops currently available. This pointed to a vicious circle of interacting reasons for growing tobacco, summed up by participants as a "seductive trap."

The overall reasoning of the participants was that the obligation to BAT (upper left quadrant) combines with the appeal of a lump-sum payment, the attractive price for the best grade of tobacco and the narrow range of cash crops currently available to keep farmers bound to the tobacco crop. The interaction of these factors in turn further undermines experimentation with new cash crops and reinforces the dominant land use. Over time, the diverse crops and technologies of farming (native seed, technical knowledge, integration with livestock, etc.) and markets for food and other cash crops have withered away. Productive uses of family labor also decline. For example, livestock management and marketing of foodstuffs no longer keep women and children employed. This means that the availability of family labor is primarily an effect of the other factors (lower right quadrant). Taken together, all of these factors combine to keep farmers dependent on an entrenched monocropping system with no process in place to identify new cash crops and innovate their way out of tobacco farming and its many negative environmental, occupational and livelihood hazards. The exercise became a consensus-building process aimed at making sense of farmer decision-making while also focussing attention on causal factors they could control (experimentation with new crops; more realistic assessments of average prices for tobacco actually paid) and that could have a positive chain effect on other related factors.

An assessment of the same question was also done separately by the research team through survey questions and responses collected in 2007 and again in 2011. The two lines of inquiry fed into each other and helped to triangulate information for a more complete view of the situation. The responses of

Table 5.2. The interaction of reasons why farmers continue to grow tobacco

Factors	Pays well	BAT obligations	Lump sum payment	Narrow range of cash crops	Uses family labor	Most grow tobacco	Cause Index Total score
Pays well	x	2	4	3	3	4	16/25
BAT obligations	3	x	4	4	3	3	17/25
Lump sum payment	4	2	x	4	3	4	17/25
Narrow range of cash crops	4	2	3	x	1	5	15/25
Uses family labor	3	0	0	2	x	3	8/25
Most grow tobacco	4	2	3	5	3	x	17/25
Effect Index Total score	18/25	8/25	14/25	18/25	13/25	19/25	90/150 (62%)

Source: Authors' field notes.

60 active tobacco farmers to an open question posed during a 2011 survey by UBINIG regarding the reasons they grow tobacco are instructive in this regard. Some 90 percent of respondents said that the lump-sum payment for their product was a major reason they continued to grow tobacco. They explained that this form of payment made it possible for them to make major purchases. They also expressed concerns over the dependency the lump-sum payment created when debts to the company were applied against the payment and led to new loans and obligations to the company. Other factors highlighted by tobacco farmers during the survey were the ease (and lack of stress) with which inputs could be acquired and the ease of selling the product to a ready buyer (noted by 55 percent of respondents). Only 37 percent of the 60 tobacco farmers interviewed mentioned the relative profitability of tobacco production as a reason for growing the crop. As we will see, below, the profitability of tobacco depends on many variables linked to the returns from land and labor within a particular household – factors that explain why this reason for growing tobacco does not stand alone in farmer decision-making. Before getting into the details of the relative economic performance of tobacco as a crop, however, the following section examines some of the tactics the tobacco companies use to pressure farmers and reinforce their dependency.

Figure 5.4. The interaction of reasons why farmers continue to grow tobacco, showing cause–effect relationships

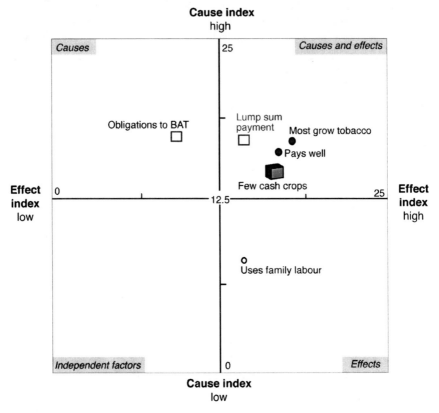

Source: Buckles 2008.

Company tactics

The causal factors driving farmers to grow tobacco interact in complex and uncertain ways with household and land tenure dynamics and the broader political and economic conditions that undermine the general development of the agricultural sector in Bangladesh. These causes and conditions are intensified by a variety of company tactics aimed at farmers and policy makers, wrapped up in the mantle of technical assistance and social corporate responsibility.

Tobacco companies such as BAT – and other national companies such as Nasir Tobacco, Akij Tobacco and Abul Khair Tobacco – emphasize the technical assistance and assurances they provide to contracted farmers. The presence of company representatives is particularly influential during pre-harvest and post-harvest periods in May and again in September and October. These seasons correspond to the yearly cyclical phenomenon of food shortages in Bangladesh known as the *mora kartik*. Thousands of rural workers and landless farmers are on the move during this period, contributing to an atmosphere of generalized anxiety in the countryside. BAT and other tobacco company staff are very active in tobacco production regions during these times, offering cash credit and renewing contracts for fertilizer and other inputs. If payments for the tobacco harvest did not meet expectations, tobacco farmers are also pressured to take advances and pay off any number of loans and debts in anticipation of a lump-sum payment and additional advances the following year.

After the tobacco crop is harvested tobacco companies also distribute seeds for sesbania bispinosa, a shrub that acts as a cover crop. The agronomic benefits of this practice are minimal, however, as the sesbania biomass is never enough to regenerate the soil or meet more than a tiny amount of the fuel requirements for curing tobacco. The tactical reason for this practice is to ensure that the land is not converted to other uses after the tobacco harvest. The cover crop occupies the land during the period between tobacco harvest and the establishment of new tobacco seedbeds. This launches farmers into a new tobacco season and creates a physical barrier to the planting of aus rice, jute and other crops that normally precede the start of the tobacco season by a few weeks.

Tobacco companies also use pricing promises to entice farmers into committing their land to a new season. The procurement price of tobacco leaves for the following season is typically set after farmers have been paid for the current season. When production declines one year, prices rise the next, sometimes dramatically, in an effort to reignite farmer excitement over the price they can anticipate the following year. These prices fluctuations from year to year are not a response to global prices for tobacco leaf but rather as a tactic to manage supply locally.

A related manipulation of the pricing system is the use of sub-categories of tobacco grades not mentioned in contracts and applied during the time of purchase to shave off the amount actually paid to farmers. This practice is widespread, but illegal in the context of a signed contract that specifies the primary grading structure and prices but does not specify the price to be paid for the additional sub-categories. For example, under grade 1, the company later introduces three sub-categories, each with a different price – if 1a is BDT 125 per kg, then 1b is BDT 123 per kg and 1c is BDT 120 per kg. Sub-grades have been created for each of the eight grades in the contracting system. Tobacco companies act as though they have the authority to modify this feature of the contract unilaterally. So far, no tobacco farmers have taken companies to court for breach of contract. However, there has been anger and protest by tobacco farmers when they experience these differences in prices. The end result can make the difference between success and failure in any particular year.

Tobacco company tactics aimed directly at farmers combine with a number of broader tactics intended to manage the public image of the tobacco companies and their relationship with local and national government agencies. The most visible of these are billboard campaigns to promote easily recognizable social issues such as biodiversity conservation, climate change adaptation, clean air and tree plantations. While these are all worthy causes, in Bangladesh there is little concrete and meaningful company action behind the campaigns. Green-washing campaigns are symbolized most cynically by the slogan *"Sobujer Somaroho"* (the abundance of green) to refer to large green fields of tobacco plants. It is misleading because tobacco cultivation is directly responsible for deforestation and environmental degradation. Another is the *"Probaho"* project which purports to help people in areas affected by arsenic in the groundwater. BAT has established 38 water treatment plants in six districts

Photograph 5.4. BAT sign promoting reforestation, with degraded tobacco field and denuded hillsides in the background

Photo credit: Daniel Buckles.

but most of these are located on fenced government premises with restricted access to the treated water. The funds invested in the treatment plants are only a fraction of what the company spends to advertise the corporate responsibility program. From this, one can conclude that the primary purpose of the billboards and related campaigns is to connect the company logo with the causes. This happens even though placing the company logo on billboards is a violation of the country's tobacco-control law, which bans advertisements by tobacco manufacturers. This kind of advertising, a longstanding practice of the tobacco industry, remains largely unchallenged in Bangladesh.

Developing a Strategy for Transition

Developing a tobacco transition strategy was not the starting point for the UBINIG research team when it first began to work with tobacco farmers. The team thought, as many studying the problem still do, that the challenge would be to come up with economically sustainable alternatives that are direct *substitutes* for tobacco. The perspective changed, however, when UBINIG began to develop and plan experiments with farmers. Initially the research team proposed to experiment on a small-scale with crops that could be substituted for tobacco on the same soils and at the same time. This included crops such as mustard seed, lentils and pulses traditionally grown during winter season in the region. Because these crops had established markets in other areas, they seemed like good candidates for field experiments and the comparative assessment of the stream of costs and benefits.

Conducting field experiments is no small matter for farmers, even on a small scale. Land and other resources dedicated to an experiment expose farmers to risks they may not be able to afford. Mindful of this, the research team decided to first engage with farmers from various parts of Bangladesh in thinking about crops of interest and what these crops bring to their farming systems. The underlying strategy was to tap into traditions of knowledge and continuous learning – what farmers actually know about crops as well as their capacity to "think outside the box." Farmers and the research team based the course of action that eventually emerged on radical innovation by farmers in the way the tobacco problem was understood.

The first step in the research process involved an assessment of the technical problems affecting tobacco lands, such as soil degradation, heavy weed burdens, the loss of animal fodders, constraints on crop rotations, a dropping water table, etc. Analysis of the interaction of these factors, using the System Dynamics method described above, produced an overall picture of the technical constraints farmers face when trying to shift out of tobacco production. It also informed decisions by farmers and the research team about what needed to improve first in order to create the possibility of more long-term change on tobacco lands.

The main conclusions from the assessment were that tobacco farming creates a downward spiral in soil and land conditions that is very difficult to reverse. On degraded soils, crops cannot initially produce enough biomass to bring about soil improvements. This in turn undermines farmers' efforts to create fodder for livestock and make full use of the potential growing season and available soil moisture. The problem is exacerbated by tough and persistent weeds such as Orobanche. The situation, farmers concluded, could only be overcome by establishing an upward spiral of gradual soil improvement through the use of diverse crops throughout the year, the introduction of leguminous crops and shrubs (that can fix atmospheric nitrogen), the production of compost with materials from various sources (including the water hyacinth and animal manure) and the gradual reintegration of livestock into the farming system. As discussed further below, this process takes time and is part of the financial cost of a transition out of tobacco farming.

The research team used this systems perspective as an input to assess the crop characteristics and cropping patterns of interest to farmers (Figure 5.5). The goal was to identify novel options for field experiments farmers had not previously considered from within their own knowledge system. Using a method adapted from Personal Construct Psychology (Chevalier and Buckles 2013), different farmer groups first identified sets of contrasting crop characteristics relevant to the transition out of tobacco production (for example, impact on soil fertility, fit within the tobacco season, market orientation, primary use, use of family labor, etc.). Each crop was then rated on each crop characteristic, using a scale of 1 to 3 represented by white, grey and black cards. Farmers gave ratings by locating each crop on a continuum of 1 to 3 between the two poles of each contrast. For example, a score of 1 if the crop helps increase soil fertility and a score of 3 if the crop needs to have fertilizer added. The score 2 was given if the crop occupied an intermediate position between the two poles. Organization of the information into a table on the ground allowed farmers to interpret the results as the exercise proceeded. For example, in Daulatpur in Kushtia participants observed that crops that increase soil fertility (left side of Figure 5.5) can also be cultivated as mixed crops and with family labor (scores also tending towards the left side of Figure 5.5).

The research teams used the data emerging from discussions with farmers to validate the assessment of major cropping patterns using a statistical technique called Principal Component Analysis. An example from Daulatpur shows the multi-dimensional relationships among the observed variables (Figure 5.6).

Figure 5.5. Crops rated on contrasting crop characteristics, using a scale of 1 to 3 between the two poles of each contrast, Daulatpur, Kushtia

Helps increase soil fertility	1 1 1 1 1 1 2 2 2 2 3 3 3	Fertilizer needs to be added
Fits within tobacco season	1 1 1 3 3 2 3 3 1 2 1 1	Straddles tobacco season
Internally oriented	1 3 3 2 3 3 1 3 2 1 1 1 3	Externally oriented
Grows well with available soil moisture	1 1 2 3 2 1 2 1 3 2 3 2 3	Water needs to be added
Primarily food/fodder crop	1 2 2 1 2 2 1 2 1 1 1 1 3	Primarily cash crop
Can be cultivated as mixed crop	1 1 2 1 2 2 1 1 2 1 3 3 3	Mono-crop only
Family labour sufficient	1 1 1 1 2 2 1 3 1 2 2 2 3	Hired labour required
Farmers can market it	1 1 1 1 2 1 1 3 1 1 1 1 3	Must be sold through broker

Tobacco
Wheat
Winter rice
Potato
Cucumber
Sugarcane
Corn
Peanut
Jute
Amaranthus
Garlic
Coriander
Mosura lentil

Photograph 5.5. Woman rating crops based on contrasting sets of crop characteristics, Daulatpur, Kushtia

Photo credit: Daniel Buckles.

Figure 5.6. Principal component analysis of the multi-dimensional relationships among crops and crop characteristics, Daulatpur, Kushtia

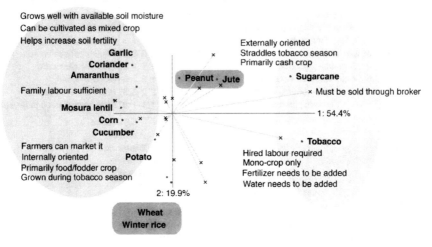

The figure shows that cropping patterns in Daulatpur are split into two competing sub-systems. One sub-system is composed of crops oriented towards local (internal) food and fodder needs (left side of Figure 5.6). These crops can be taken to market directly by farmers and managed using family labor. The crops tend to be planted and harvested during the same season as tobacco, lend themselves to mixed cropping, grow well with available soil moisture and help increase soil fertility. Examples are garlic, coriander, amaranths and *mosura dal* (a pulse). The second sub-system consists of crops oriented towards an external market. These crops must be sold through brokers, mill owners or company buyers before they get to market. They tend to be grown as a monocrop for cash and require additional inputs such as water, fertilizers and hired labor. Tobacco and sugarcane are examples of crops that combine these characteristics.

Tobacco farmers participating in the analysis, which was repeated several times in Kushtia, Bandarban and Cox's Bazar, explained that by combining both sub-systems farmers can grow food and fodder for local needs and secure external income. They felt that this dual strategy was necessary but also problematic because they compete directly with each other for land and other resources. The dual strategy forces farmers to choose between the two sub-systems or split their land into two separate blocks. This observation is consistent with findings regarding farmer profiles reported above: the landless and smallholders choose to specialize in tobacco production while larger farmers grow food-fodder crops and tobacco on separate parcels of land. These conflicting cropping systems limit the scope for a transition out of tobacco production, a dilemma that cannot be resolved from within the set of strategies currently available to tobacco farmers.

The analysis did stimulate new ideas about how to go beyond the dual strategy in the current farming system. Figure 5.6 also shows that some crops stand apart from all others in that they combine crop characteristics quite differently. Crops such as wheat, rice and to some extent potato are grown in a way similar to monocrops, where external inputs are needed. Unlike the externally oriented crops, however, they contribute strongly to existing local food and fodder systems and can be marketed by farmers themselves. These crops usually occupy the land at the same time as tobacco. A second group of crops (jute and peanuts, for instance) combine features in novel ways as well. They are like local food and fodder crops in many respects but also have well-established external markets. Furthermore, they do not compete directly with tobacco but rather straddle the tobacco season, starting either before or after tobacco is in the ground. This observation led to thinking by farmers about unusual combinations of desirable crop characteristics and created an important learning opportunity for the farmer participants and the research team.

Discussion of these novel combinations of crop characteristics generated a lot of excitement among participating farmers because it provoked a shift in thinking about what is possible over the entire agricultural year. Rather than trying to find the perfect crop to substitute for tobacco during the same season, farmers and the research team started to think about ways to support a gradual transition into different cropping patterns. In practical terms, the transition needed to begin in the season before tobacco is grown. Planting of new crop combinations would allow for some improvements in soil quality and reduction of the weed burden and potentially initiate a stream of financial benefits that could continue throughout the year.

One of the women farmers in the group, Sheuli Begum, gave shape to this new thinking with an example. She explained that while she was going to the local market in search of new cash crops, she saw a spice that she and other rural women buy regularly. It contained seeds of three different plants not currently grown in her region. She sprouted and planted all three seeds, and then chose fenugreek (*methi*), which she believed could be easily grown in a mixed cropping system. It needed to be planted before the tobacco season, and would help create some cash income at a time when she would be tending to other crops that would mature later on. Inspired by this idea, participants decided to also search local and regional markets for products that combine crops and crop characteristics in novel ways – crops where there is a demand in markets they can access themselves (either locally or regionally), that can be grown in mixed cropping systems that improve the soil or that straddle the beginning or end of the tobacco season (and therefore provide a transitional stream of income).

The distinction between *substitute crops* and *transition cropping patterns* had not occurred earlier to farmers or the research team and provided a new lens to reflect on strategies for shifting out of tobacco production. It moved thinking by the groups of farmers beyond current local knowledge and conceptual categories into a new space of innovative thinking and experimentation. Various promising crop combinations and cropping patterns emerged, drawing on results. For example, farmers from Cox's Bazar argued that potato, French bean and felon bean would combine well in the winter season, and support a cropping pattern involving aus and *aman* rice later in the year. Potato, maize and lentil combinations seemed promising to farmers in Kushtia, also combined with jute and off-season rice. Farmers in Bandarban initially proposed to combine potato, tomato and felon but later decided to add coriander to the mix, allowing for a better balance in terms of harvest time and seeding of later food and cash crops. Each set of crops that farmers discussed combined desirable crop characteristics and interacted with other cropping patterns in positive and novel ways.

Photograph 5.6. Woman collecting radish and potato from her mixed cropping field converted from tobacco

Photo credit: Abdul Zabbar.

In the weeks following these assessments, farmers in each region converged in their thinking about the best decisions. They also requested support from the research team for access to seeds of the selected crop combinations. The research team then acquired seeds for the candidate crops from Nayakrishi Andolon, a network of farmer-run seed centers located in different parts of the country. These centers have a long history of testing and selecting local varieties of diverse crops of interest to farmers and developing detailed knowledge regarding their characteristics, management requirements and seed production procedures.[5]

The research team made a fixed amount of seed available to farmers committed to establishing field experiments on land where they had previously grown tobacco. While many opted to test the recommended combinations in their entirety, others adopted slightly different combinations in light of their own judgments and preferences. Over a period of two years (2010–2011) some 365 farmers set up field experiments, generating a wealth of new experience and stimulating a concerted effort to innovate and evaluate outcomes.

Assessing economic viability

Various farmer- and researcher-led studies assessed the economic viability of the experimental results. The most systematic and comprehensive of these was a cost–benefit analysis of the financial performance of new crop combinations compared to tobacco production, using three years of survey data on actual costs, yields and crop prices (UBINIG 2007, 2008, 2011; Uddin Molla 2011).[6] The results from 2011 (Table 5.3 and Table 5.4) show that gross returns on tobacco and mixed crop combinations are similar but that labor costs and the cost of purchased inputs (fertilizers, pesticides, irrigation water) were much higher in the tobacco production system. On average, 415 person-days per hectare were used in tobacco while in mixed cropping systems 231 person-days, a difference of 88 percent. The wage rate paid to workers in tobacco fields was also higher, 21 percent higher on average, due to the continuous and demanding nature of the work. Fertilizer costs were not only higher in the tobacco production system but also excessive and imbalanced when compared to national recommendations.[7] Irrigation

---

5   For details on the Nayakrishi Andolon seed system, see www.ubinig.org.

6   The 2007 and 2008 surveys involved a purposive sampling technique to include 24 tobacco farmers and 36 farmers experimenting with new crop combinations. In 2011, a random sample was developed from lists of tobacco farmers and farmers using new and established combinations of alternative crops and included 60 tobacco farmers and 90 other farmers. The questionnaires were tested and enumerators trained prior to conducting interviews, which occurred shortly after the winter (rabi) harvest.

7   Average doses of fertilizer in tobacco fields in 2011 were: Urea (575 kg/ha), TSP (366 kg/ha), MoP (35 kg/ha), SoP (235 kg/ha), DAP (44 kg/ha) and Znso4 (57 kg/ha). These greatly exceed recommendations and do not reflect an appropriate balance of nutrients for optimal plant growth.

Table 5.3. Costs and benefits per ha of tobacco production at different locations in Bangladesh, 2011

Items	Cost BDT / ha*			
	Kushtia	Cox's bazar	Bandarban	Average
Human labor:				
Family	12,118	21,659	6,516	
Hired	30,806	53,879	72,943	
Total	42,924	75,538	79,459	65,974
Land preparation:				
Owned plough equipment	–	–	–	
Hired plough equipment	3,790	6,108	6,220	
Total	3,790	6,108	6,220	5,373
Seed/seedlings:				
Owned	–	–	–	
Purchased	1,890	2,161	2,657	
Total	1,890	2,161	2,657	2,236
Fertilizers	42,228	44,344	36,927	41,166
Insecticides / pesticides	4,854	12,142	10,146	9,047
Irrigation	4,564	9,645	9,417	7,875
Manure:				
Owned	117	415	803	
Purchased	1,297	712	–	
Total	1,414	1,127	803	1,115
Drying/Curing:				
Owned fuel	1,102	2,121	–	
Purchased fuel	39,070	45,945	42,596	
Total	40,172	48,066	42,596	43,611
Others (transportation, stick, rope, medicine, etc.)	6,361	8,818	7,078	7,419
Interest on working capital @ 7%	5,507	7,503	7,676	6,895
Rental cost of land (for crop season only)	25,319	32,419	36,021	31,253
Total cost (TC)	179,023	247,871	239,000	221,965
Total cash cost (TCC)	134,860	183,754	187,984	168,866
Yield of crops (kg/ha)	1,775	2,220	1,898	1,964
Price of crops (BDT/kg)	102	120	125	116

(*Continued*)

Table 5.3. Continued

Items	Cost BDT / ha*			
	Kushtia	Cox's bazar	Bandarban	Average
Gross Return (BDT/ha)				
Value of Crops	181,050	266,400	237,250	
Value of by-products	3,445	6,989	7,927	
Total	184,495	273,389	245,177	234,354
Net Profit (BDT/ha)				
Full cost basis	5,472	25,518	6,177	12,389
Cash cost basis	49,635	89,635	57,193	65,488
Return per BDT Investment				
Full cost basis	1.03	1.10	1.03	1.05
Cash cost basis	1.37	1.49	1.30	1.39

* Exchange rate 76 BDT to 1 USD.

water was also needed on average six times in a tobacco season compared to three applications in the tested mixed cropping systems. These add up to costs per hectare of tobacco 119 percent higher on average compared to mixed cropping systems.

As a result of these higher costs, net profit was much lower in tobacco. The rate of return on investment was also much lower in tobacco: 1.05 compared to 2.47 per BDT invested in mixed cropping systems on a full cost basis (costs, including in-kind labor and other in-kind costs) and 1.39 compared to 5.40 per BDT invested in mixed cropping systems on a cash cost basis (costs, including only cash expenditures). The comparison helps to explain why the lower productivity of tobacco farming has not been immediately evident to tobacco farmers: cash advances to cover costs and the high prices and large lump-sum payment for processed tobacco leaves obscure the calculation of total costs and final profits. The contracting system hides deep flaws in the business case for tobacco.

A full-cost accounting of economic performance from a farmer perspective points to an even sharper contrast between tobacco and other crops of interest to farmers. In a 2007 assessment, farmers in several villages involved in the experiments generated categories of gains and losses relevant to their farming practice and indicators for each category.[8] These were elicited,

---

8   For a description of the method called "Interests" used to do the assessment, see Chevalier and Buckles 2013.

Table 5.4. Costs and benefits per ha of different food and fodder systems at different locations in Bangladesh, 2011

Items	Cost BDT / ha*			
	Kushtia	Cox's bazar	Bandarban	Average
	Potato+maize+lentil+coriander	Potato+french bean+felon	Potato+french bean+felon	
Human labor:				
Family	15,222	15,750	17,850	
Hired	15,649	12,000	8,700	
Total	30,831	27,750	26,550	28,377
Land preparation:				
Owned plough equipment	–	2,733	4,281	
Purchased plough equipment	3,255	3,549	1,997	
Total	3,255	6,282	6,278	5,272
Seed:				
Owned	1,938	1,157	1,174	
Purchased	38,182	24,095	25,603	
Total	40,120	25,252	26,770	30,714
Manure:				
Owned	4,283	4,672	6,168	
Purchased	–	1,546	189	
Total	4,283	6,218	6,357	5,619
Irrigation:				
Owned	–	–	1,009	
Purchased	2,659	4,081	3,025	
Total	2,659	4,081	4,034	3,591

Others (if any)	651	301	512	488
Interest on working capital (at 7%)	1,057	798	701	852
Rental value of land (for crop season only)	21,188	28,391	29,933	26,504
Total cost (TC)	103,688	99,073	1,01,142	101,301
Total cash cost (TCC)	60,396	45,572	40,026	48,665
Yield of crops (kg/ha)	15,728 (potatoes); 2845 (maize); 116 (lentil); 81 (coriander)	13,457 (potatoes); 567 (french beans); 303 (felon)	13,560 (potatoes); 616 (french bean); 328 (felon)	–
Price of crops (BDT/kg)	11.0 (potatoes); 11.7 (maize); 72.5 (lentil); 67.0 (coriander)	18.0 (potatoes); 45.0 (french bean); 45.0 (felon)	18.0 (potatoes); 43.0 (french bean); 42.5 (felon)	–
Gross return (BDT/ha):				
Value of crops	209,767	242,226	284,508	
Value of by-products	7,160	3,946	3,763	
Total	216,927	246,172	288,271	250,457
Net profit (BDT/ha):				
Full cost basis	113,239	147,099	187,129	149,156
Cash cost basis	156,531	200,600	248,245	201,792
Return per BDT Investment:				
Full cost basis	2.09	2.48	2.85	2.47
Cash cost basis	3.59	5.40	7.20	5.40

* Exchange rate 76 BDT to 1 USD.

sorted and piled by participants into five categories, each with a different weight reflecting their relative importance (Table 5.5). A variable scale (for example, from +10 to –10 for Yield, the most important category of gains and losses) expressed the weight of each category. Farmers scored the actual economic performance of their individual cropping system in a particular year and compared scores among themselves. This generated a great deal of discussion about the various factors that go into calculating economic performance and created new insights into the complex process of farmer decision-making.

The farmer-generated categories of gains and losses, and indicators for each, were included in the 2007 survey to support a broader consultation regarding farmer perspectives on economic performance. Results showed that about 20 percent of the tobacco farmers surveyed experienced severe difficulties repaying their loans (Monetary returns) and 30 percent did not have enough food to meet their needs for three months or more (Food independence). Only a few farmers testing alternative combinations experienced food shortages and none had problems repaying loans. For one criterion, Yield, the alternative crop combinations did not do as well as tobacco production – yield fell below expectations for almost half of the farmers testing new cropping patterns that year. Only one-third of tobacco farmers experienced a similar level of disappointment in the expected yield of tobacco leaf. Farmers explained that they did not achieve expected yields for transition crops because their soils had been degraded after many years of tobacco cultivation. This finding underlined the need to focus the transition strategy on steps to incrementally improve soil quality.

While interesting in many ways, the farmer-generated categories and indicators were not used in the 2008 and 2011 economic performance surveys. The research team decided that conventional ways of measuring financial returns on land and labor costs were good enough for discussions with policy makers, the main audience for the survey results. Instead, the team decided to create a tool based on the criteria and indicators that men and women farmers could use independently to assess their own economic performance. Women in farming households formed separate groups to develop their own collective position on what criteria to include in their analysis and rate the performance of particular crop combinations. Men in farming households did the same. These discussions allowed for a grounded and gender-sensitive discussion of how household and field circumstances affect economic performance, and different points of view regarding what goes into making "the best decision." Farmers said that the exercise helped them think through the various considerations that implicitly go into an overall assessment of a farming season. "I can now assess myself," said one

Table 5.5. Farmer-generated categories of gains and losses and sample scores from two farmers (tobacco and mixed crop)

Categories of gains and losses	Graduated indicators	Scale	Net Gain or Loss	
			Tobacco farmer	Mixed crop farmer
Yield	Yield much higher than expectation (+10) Yield more than expectation (+5) Expected yield achieved (0) Yield not up to expectation (−5) Yield much lower than expectation (−10)	+10 to −10	+5	−4
Monetary returns	Money available for purchase of cow (+6) Some money remaining in hand after the repayment of loan (+3) Loan has been repaid somehow (0) Need loan for the next crop (−3) Need additional loan to meet needs (−6)	+6 to −6	−6	+3
Happiness	Very happy (+4) Happy (+2) No worry and anxiety (0) Lack of peace in family life (−2) Disagreement between wife and husband (−4)	+4 to −4	−4	+4
Product quality	Very good quality (+5) Good quality (+2.5) Average quality (0) Low-quality (−2.5) Product half ripe, undeveloped, rotten (−5)	−6 to +6	+3	+3
Food independence	Surplus food for sharing among relations after meeting family needs (+8) Enough food throughout the year (+4) Minimum food independence ensured (0) Lack of food for one month (−4) Lack of minimum food for 3 months (−8)	−8 to +8	−4	+6
Overall		−34 to +34	−6	+12

farmer in reference to how he was now thinking about the strengths and weaknesses of his season and where he needed to pay more attention in the future. The process generated a great deal of confidence among farmers that breaking the bonds of tobacco farming was in their own economic interests.

Photograph 5.7. Farmers assessing crop combinations using criteria and indicators they developed

Photo credit: Daniel Buckles.

Scaling Up the Transition Strategy

The complete cycle of diagnosis, innovation, experimentation and grounded evaluation resulted in economically viable transition strategies adapted to two different tobacco-growing regions. After two years of research with farmers UBINIG could confidently conclude that tobacco farmers felt more optimistic about the possibility of a transition than was initially the case. Before the experiments, tobacco farmers had depended on whatever price the companies offered for the tobacco leaves. Even if farmers were not satisfied with the prices, they had to continue. After the experiments showing them other possibilities, farmers could imagine a transition out of tobacco. For every farmer that joined the research there were many more expressing interest, in the same villages as well as in neighbouring villages. This was encouraging.

Some key questions remained unanswered, however. The research team recognized that to sustain the gains made we needed to engage government officials, scientists and the public in supporting an active and orderly transition out of tobacco production. This prompted a new cycle of participatory action research beyond the farm and village level. It engaged a national farmers' movement (the Nayakrishi Andolon), local and regional markets and tobacco-control organizations and institutions within and outside of government.

Extending farmer-based assistance

The immediate and fundamental demand of farmers wanting to shift out of tobacco production was access to locally adapted seeds and information about suitable crop combinations. To meet this need, the research team turned again to the Nayakrishi Andolon, an established farmers' movement in Bangladesh comprised of more than 300,000 households in 19 districts of the country. For many years UBINIG and the Nayakrishi Andolon have worked together to develop the Nayakrishi Seed Network, a well-structured system of seed management at the household, village and regional levels. Women play a significant role in the network, preserving collectively some 3,000 varieties of rice and hundreds of varieties of Bangladeshi vegetables, pulses and other native plants. Seeds managed at the household level are represented collectively at the village level in a Nayakrishi Seed Hut managed by village women. The women replant, regenerate, conserve and sell these seeds to cover costs. Samples of seeds are also collected in a district- or region-level Seed Wealth Centre, where farmers from anywhere can directly access seeds of interest or be directed to the village level source.

Building on several decades of famer-based extension experience, the Nayakrishi Andolon developed a farmer exchange process adapted to the needs of tobacco farmers. Between 2009 and 2011, some 220 tobacco farmers visited Nayakrishi Seed Wealth Centers and Seed Huts, learning about the identification of crop varieties, seed germination and viability and management of stored seed. They also observed and discussed with experienced Nayakrishi farmers and UBINIG staff the farming practices best suited to supporting a transition out of tobacco production, including mixed cropping, crop rotation, the making of compost and management of crop diversity. Particular attention was given to the calculation of input costs and the broad range of economic benefits from mixed crop production, taking into account factors beyond the yield of a single crop grown in a monoculture. This helped farmers situate the transition out of tobacco production in broader terms, including how crops are linked to livestock and water use, the costs of chemical dependence and the many incremental gains that stem from reduced household expenditures and a small but steady stream of income. The exposure and training process for each group of farmers lasted three to five days, often culminating in decisions by farmers to establish Seed Huts in their own communities. Importantly, 38 percent of participants were women. Given their role in Bangladeshi farming communities as seed keepers, this level of participation increased the chances of successful establishment of the Seed Huts and follow-through on household decisions regarding what seed to use.

Following the farmer exchange, Nayakrishi Andolon and UBINIG offered tobacco farmers seed of local *aman* rice varieties they could plant immediately after the tobacco harvest (*Kharif-1* season). The short-maturing varieties of *aman* rice were of particular interest to farmers since they could be harvested in time for winter (*rabi*) crop cultivation. Seeds of other short-duration crops that could be easily sold at market were also offered, including spinach, radish, coriander and amaranth. These pre-empted the short fallow period and preparation of tobacco seedbeds that were part of the normal tobacco cycle. Farmers accepting the seed were asked to commit to using the ecological farming practices learned from Nayakrishi farmers and to return to the Nayakrishi Seed Network the same amount of seed received as support. Similar arrangements were made for the winter cropping season, using seed combinations unique to each region (as identified in the tables above). In both study regions, several Nayakrishi Seed Huts were established at the village level in communities dominated by tobacco production, thereby providing tobacco farmers not directly involved in the research with direct access to seeds and related knowledge of local farmers.

Between 2009 and 2011, 411 different farm households – 144 in Kushtia and 267 combined in Bandarban and Cox's Bazaar – used the transition seeds offered. Typically, they did so in the ways recommended, including adoption of the ecological farming practices of the Nayakrishi Andolon. Because each household was only offered seed once, in subsequent years farmers accessed seed from their own supply or by buying seed from a Nayakrishi Seed Hut or other local source. Furthermore, virtually all of the households involved returned after the harvest the same quantity of seed they had received. This helped with seed distribution the following year.

When surveyed in 2012, none of the 365 former tobacco farmers participating in the experiments had returned to tobacco production. For them, the transition out of tobacco was complete and sustained without additional technical assistance or seed inputs. While not enough to ensure continuous innovation and development of new seed sources, the intervention established the beginnings of a local and regional seed system.

Strengthening market opportunities

The farmers involved in the transition frequently noted that their calculations of profitability now took into account the full range of costs and benefits realized over the course of a year, rather than simply during one cropping season. They said this was a key reason for not going back to tobacco production. They also struggled, however, with marketing their crops. After many years of tobacco farming, local and regional markets for food crops and other potential outputs from farmers' fields had weakened or died away completely. District-level markets were often far away from tobacco-growing areas and prices for other products low at the selling point. This created a serious hurdle compared to marketing of tobacco leaves at nearby company buying houses.

To respond to this constraint, UBINIG began marketing efforts that involved negotiating arrangements with officials to establish a separate outlet in the local and regional markets for food and other crops grown using ecological methods. This reduced the individual cost for market-stall rentals, facilitated direct sales by farmers to buyers and created publicity for products marketed as free of pesticides. UBINIG also encouraged farmers to transport their goods to market collectively, and provided, whenever possible, access to common transportation and a local warehouse for cold storage. Demand for transportation and storage far outweighed UBINIG's resources, however, leaving many former tobacco farmers on their own in this regard. The research team was, nevertheless, able to collate and share information on prices for a wide range of products at village and regional markets and wholesale depots so that farmers could make independent decisions about where to sell their products. It provided assistance as well to the calculation of what constitutes a fair price for specific products, based on detailed estimates of the costs of production.

Bolstering national food security

While meaningful for the households involved, UBINIG efforts to strengthen market opportunities could not approach what is needed to fully counter the tobacco company's market supports. They show, however, that very modest investments in improving farmer access to local seed and related knowledge, transportation, storage facilities and market information can make a difference for farmers and contribute to local and regional food markets. In all, more than 80 hectares of land previously under tobacco was directly converted to food and other cash crops during the project period. Much of this land is suitable for a wide range of food crops, including vegetables, pulses, spices, edible oil seeds, jute, rice and wheat.

Table 5.6 provides an estimate of the market value of tobacco and mixed crop combinations for two unions in the study.[9] It shows a plausible lower limit to the economic value of converting tobacco lands in these unions to food production. Upper limit contributions to food production would include as well the potential market value of crops grown on the same land over the entire year.

Table 5.6. Estimated market value of tobacco and mixed crop combinations (winter season only)

Union*	Land under tobacco production (ha)	Market value of tobacco grown in 2011 (millions BDT)	Mixed crop combination	Potential market value of mixed crop combination (millions BDT)
Lama, Bandarban District	4,083	1,001	Potato + french bean + coriander + sweet gourd	1,177
Daulatpur, Kushtia District	8,093	1,493	Potato + maize + garlic + corn + coriander	1,755

* Estimates from each union are based on average yield and average prices paid for tobacco and food products. Exchange rate, 8 June 2011 was 1 USD equivalent to 72.5 BDT.
Source: Authors' 2011 survey data.

The potential contribution to food production of a complete transition out of tobacco is potentially strategic to national food security and employment objectives (Rahman 2011). While tobacco currently occupies slightly less than one percent of all arable land in Bangladesh, much of the tobacco land has the potential for two or even three cropping seasons per year when used in mixed cropping systems. These high productivity lands are needed to address downward national trends in the production of pulses, oil seeds, spices, condiments and winter vegetables that are currently the chief sources of protein, minerals and vitamins for most of the population (Rahman and Khan 2005; Brammer 1997). Converting all of the double- and triple-cropped land where tobacco currently grows back into food production would reduce the volume of pulses imported into Bangladesh, estimated by Mishra and Hossain (2005) to be 30 percent of the effective demand. It could also help to reduce

---

9   Unions are the lowest tier of regional administration in Bangladesh, below the *upazila*. These two unions provided the most complete and reliable data on the total land area under tobacco production in 2011 for the study area.

foreign exchange losses, which in 2011 amounted to about BDT 1,200 billion for the import of 260 million tons of food.

Areas such as Chakaria in Cox's Bazar and *upazilas* in Bandarban along the Matamuhuri River – both of which are prime targets for the expansion of tobacco in Bangladesh – are particularly rich and productive environments capable of becoming significant food baskets for populations in southeastern Bangladesh. Kushtia, which once served this function for central Bangladesh, could once again contribute significantly to food security by converting lands currently dedicated to tobacco into fields of food and other crops of value to Bangladeshi society.[10]

From Field to Policy Debates

The action-research process launched with farmers in Bangladesh shows that modest investments in scaling up farmer-based extension services and strengthening market opportunities in tobacco-growing regions can create conditions for a successful transition out of tobacco production. To reach larger numbers of tobacco farmers, however, and to sustain the transition, policy action is needed at the regional and national levels.

Mobilizing the political will to support a transition out of tobacco and into food production is a challenge because the strategy comes face-to-face with the powerful tobacco lobby in Bangladesh and various structural challenges in the government of Bangladesh. The policy environment is complex and involves many government ministries with policies and responsibilities relevant to the challenge:

- The Ministry of Health has the lead responsibility for coordinating amendments to and implementation of the Tobacco Control Law, in coordination with the Ministry of Home Affairs and the Justice Ministry. While the health costs of tobacco-related illnesses in Bangladesh have been well documented (MOHFW 2009; WHO 2007; Efroymson and Ahmed 2001) the Ministry has also recently recognized the occupational health hazards associated with tobacco production, especially for women and children working in kilns. It has begun to formulate a response, in consultation with the Ministry of Women and Child Affairs, the Cultural Ministry and the Ministry of Religious Affairs.

---

10 Roy et al. (2012, 316) make a parallel argument for employment impacts of reduced tobacco use in Bangladesh. They estimate that a shift in bidi expenditures among the poor could create "a large number of higher value, healthier and better remunerated jobs [...] completely offsetting any job losses in the tobacco sector."

- The National Board of Revenue manages the Value Added Tax (VAT) paid by both national and international tobacco companies. In 2009, BAT paid about BDT 37.5 billion in taxes into the national coffers, one of the largest single contributors in the country (Choudhury 2010). Breaking the government's dependence on this revenue source would require unprecedented political will and financial planning for a fiscal transition. Nevertheless, research shows that changes in taxes on cigarettes and bidis could lead to significant reductions in premature deaths while at the same time increasing excise revenues (Barkat et al. 2012).
- The Ministry of Agriculture, and in particular the Department of Agricultural Extension, is mandated to provide farmers with technical assistance, inputs and marketing support for crop production. Existing policies would allow for the development of special extension programs targeting tobacco-growing regions. For example, the government recently launched a program to reduce tobacco cultivation by expanding cotton cultivation (Uddin Bhuyan 2012). The ministry also suspended subsidies on fertilizers used for tobacco production after realizing that it was drawing away support needed for vital *boro* rice production. Tobacco companies quickly adjusted to this change in policy, however, by encouraging tobacco farmers to purchase their subsidized urea at a time of year normally associated with *boro* rice production. These tactics have corrupted not only the policy but also the dealers and Ministry field officers responsible for dispersing the input, while also putting extra monitoring demands on the Ministry.
- The Bangladesh Bank, the monetary and financial regulator for Bangladesh, issued a circular on 18 April 2011 ordering all scheduled commercial banks to suspend the practice of granting loans to individuals and companies for the purpose of tobacco farming. The circular cited the threats to public health from the use of tobacco products but also mentions the impacts of tobacco farming on the food crisis in Bangladesh and on the environment. It also instructed banks to close previously disbursed loans for tobacco production on schedule and to not renew or extend them for any reason.
- The Ministry of Industries, which for years has been providing chemical fertilizers to tobacco companies at a subsidized rate, has recently begun to reduce the support. The scale of the reduction is still limited, however.
- The Ministry of Land is responsible for *khas* lands in Bangladesh. These lands are mandated for use by the poor and landless. Local administrative officers in Bandarban have made use of existing administrative procedures within the Ministry to ban the use of *khas* lands for tobacco cultivation, ordering the destruction of the crop on *khas* lands located on the banks of the Matamuhuri River.

- The Ministry of Commerce monitors the export of tobacco leaves and regulates the trade licenses of the tobacco companies. It is also responsible for monitoring imports and estimating requirements for essential food items in light of the National Food Policy and the Agriculture and Nutrition Policy of the government. This puts the Ministry in a position to help plan the conversion of tobacco lands needed to address strategic food gaps such as higher imports of pulses, oilseeds and rice. There are compelling reasons to do so. Garment workers, for example, represent 40 percent of the nation's working population and contribute 80 percent of the country's export earnings. Ensuring their food security is a vital national priority.
- The Ministry of Food and Disaster Management monitors food deficiency for the country as a whole and at the sub-national level. It can support policies aimed at shifting tobacco out of lands suitable for food crops and restricting the expansion of tobacco production into new areas.
- The Ministry of Labor is responsible for the enforcement of laws regulating the use of child labor – laws it can use to prohibit the use of children in certain activities such as tending fires in tobacco kilns. It can also monitor the occupational hazards that the female labor force faces during different stages of tobacco cultivation.
- The Ministry of Education is responsible for monitoring school attendance and can use its influence to address absenteeism in tobacco-growing areas. Secondary School Exams routinely take place during the tobacco-curing season, forcing children to drop out due to the demand for their labor. So far, the Ministry of Education has not taken any action in this regard.
- The Ministry of Environment and Forests is responsible for protecting the environment and could take action against industries that cause deforestation. It is also responsible for protecting reserve forests. It can use these powers to prohibit the cutting of trees for use in the tobacco industry. It could restrict or even ban tobacco production in areas such as the Chittagong Hill Tracts, which are threatened by deforestation. It can also, by virtue of its responsibilities for air and water quality, prohibit the construction and operation of tobacco kilns within 500 meters of schools and other public facilities and regulate the use of pesticides and fertilizers for tobacco production near a watercourse. The Ministry has not yet acted on these areas of responsibility. On the contrary, it has actively participated in tree plantation programs sponsored by BAT and denied claims that tobacco cultivation contributes to deforestation and other environmental hazards.

This complex institutional environment provides numerous avenues for the tobacco industry lobby to delay action and undermine important initiatives. For example, in March 2013, two key provisions in the Smoking and Tobacco

Products Usage (Control) (Amendment) Bill were struck from the final draft submitted to a standing committee for review prior to debate and decision by Parliament. One would have disallowed the designation of smoking areas in public spaces. The other would have banned the provision of any kind of government subsidy to tobacco cultivation. No ministry would take responsibility for having objected to these provisions. Instead, they cited unspecified parliamentarians and pressures by the tobacco industry on the Finance Ministry as the reasons for their failure to proceed (Kalam Azad 2013).

An effective policy to curb tobacco cultivation and encourage a transition faces many political barriers. The political discourse is changing, however, as government officials in Bangladesh begin to publicly recognize important messages from research on the topic: tobacco cultivation has many occupational health hazards, it destroys soils and forest resources and it diverts prime agricultural land away from essential food production and food security objectives. The actions of the Bangladesh Bank to suspend loans for tobacco production are particularly encouraging in this regard and suggest that there are actors in the political sphere with the integrity to resist industry pressure.

UBINIG and the Nayakrishi Andolon have actively campaigned in Bangladesh on these issues since 2006, contributing in various ways to raising public awareness and generating research results for use in the public debate. These efforts have included meetings with bank officials providing government loans for tobacco production and with civil authorities and the military charged with protecting the reserve forests of the Chittagong Hill Tracts. The organization has also worked with local government officials unaware of the illegal occupation of *khas* land for tobacco production, with Ministry of Health officials concerning the health impacts of tobacco farming and with Ministry of Agriculture officials responsible for agricultural extension and food security programs. Workshops organized by UBINIG provided Bangladeshi scientists and government officials with opportunities to discuss the challenges and options for transitions out of tobacco cultivation and to use their expert knowledge in support of planning for transition.

UBINIG has also joined with established anti-tobacco organizations in Bangladesh to celebrate World No Tobacco Day (31 May), World Environment Day (5 June) and World Food Day (16 October) with public campaigns and marches. UBINIG helped with the formation in 2010 of the Anti-Tobacco Women's Alliance (known in Bangla as *Tabinaj*) as an adjunct to the Bangladesh Anti-Tobacco Alliance (BATA). Another active alliance consists of journalists in the Anti-Tobacco Media Alliance (ATMA) who are now reporting in print and electronic media about the harmful effects of tobacco production and consumption. Over time, these movements have formed close alliances by

Photograph 5.8. Farmers and activists raising awareness about the harmful effects of tobacco farming on the occasion of World Food Day, Dhaka, 2010

Photo credit: Abdul Zabbar.

bringing together the concerns about the health impacts of smoking with recognition of the many negative forces created by tobacco cultivation. Jointly and independently, the various actors are doing what they can to articulate a coherent and integrated approach to the control of tobacco production and use in Bangladesh, including support for an orderly exit from tobacco production by tobacco farmers.

These campaigns and the action research described in this chapter also provided journalists in Bangladesh with access to new information and analysis of the under-reported issues of tobacco production. In 2010, two local journalists acted independently on these concerns by filing a public interest litigation case in the District Court of Bandarban. The submission called for a ban on the promotion of tobacco cultivation by tobacco companies due to its negative environmental and social impacts. The court judge granted a temporary injunction against tobacco farming and instructed police chiefs, forest and agriculture officers and officials of three tobacco companies active in the area to explain why tobacco kilns and warehouses should not be relocated outside the district. Ultimately, the court ruled to limit the amount of land in the Bandarban District under tobacco cultivation to 1,000 ha, down dramatically from the 10,000 ha under cultivation at the time. While still not fully implemented, the ruling and the public debate about it have created awareness among tobacco farmers, prompting many to stop cultivating tobacco. It remains a milestone achievement within the tobacco-control movement in Bangladesh.

Conclusions

Action research on tobacco transition strategies for Bangladesh focused initially on engaging farmers, developing a detailed understanding of the constraints they face and jointly creating a grounded strategy for transition. It built on farmers' own knowledge and their capacity to innovate, experiment and evaluate new information from various sources, including from agricultural scientists. Importantly, the distinction between *substitute crops* and *transition cropping patterns* that emerged from the analysis shifted thinking from the search for a perfect alternative to the development of a dynamic transition strategy making full use of the annual cropping cycle and plant genetic resources still available in Bangladesh. The process went beyond what farmers had traditionally grown in their respective regions and provided farmers a way to overcome the soil degradation and weed infestation left by years of tobacco production. Regionally adapted strategies also met the test of economic viability from various points of view, confirming that higher rates of return on land, labor and financial resources are available to farmers who can access seed, related knowledge and markets for food and other crops. This understanding modified in a meaningful way farmer economic calculations of what land and labor uses are possible, and in their best interests.

Efforts to scale up the transition strategy, and evidence that the farmers involved have not returned to tobacco farming, also show that modest investments in transitions can contribute directly to national food security objectives. This success demonstrates the broader reasons why the Government of Bangladesh would want to support an orderly transition out of tobacco cultivation and actively regulate against the expansion of tobacco growing in food-producing areas. Simply, the obligation of the government to ensure food for its urban and rural population is paramount. Given this imperative, Bangladesh cannot afford to allow tobacco companies to continue to divert scarce arable land away from food production. This security threat calls for direct action on the part of government and a political movement to press in this direction.

The challenge, while significant, is not insurmountable even for a country with limited financial resources. The key requirement is the political will to empower and compel the various ministries involved to coordinate and implement existing policies that would restrict the expansion of tobacco farming into new areas, eliminate public support for tobacco growing and make targeted investments in development of the agricultural sector. Many supportive actions by ministries can be taken within the scope of existing policies, rather than through comprehensive policy reform. Cooperation between the Ministry of Agriculture, the Bank of Bangladesh and the Ministry of Health in particular could create a geographically focused program of strategic supports to facilitate access to locally adapted and diverse agricultural seed and related knowledge and investments in cold storage, essential transportation and market organization. The careful timing of these supports, and sustained and sensitive commitment to facilitating peoples' own analysis and action, are also important. This level of mobilization and coordination is well within the current capacity of the government and is consistent with a vision for Bangladesh that values the contribution of food-producing communities to the economic and food security interests of the nation.

References

Abu-Irmaileh, B. E. and Ricardo Labrada. nd. "The Problem of *Orobanche* spp in Africa and Near East." FAO. Online: http://www.fao.org/agriculture/crops/core-themes/theme/biodiversity/weeds/issues/oro/en/ (accessed 1 December 2012).

Akhter, F., R. Haque Tito, Md. Begum Nargis and M. Islam Prince. 2012. *Tamaker Sringkhol theke mukti /Freedom from the tobacco chain/*. [Bangla] Dhaka, Bangladesh: Narigrantha Prabartana.

Bala, B. K. 2010. "Management of Agricultural Systems of the Upland of Chittagong Hill Tracts for Sustainable Food Security." Report published by National Food Policy Capacity Strengthening Programme, Dhaka, 156 pp.

Bangladesh Bureau of Statistics. 2010. "Statistical Pocket Book of Bangladesh." Online: http://www.bbs.gov.bd/PageWebMenuContent.aspx?MenuKey=117 (accessed 29 December 2013).

Barkat, A., A. Chowdhury, N. Nargis, M. Rahman, Pk. A. Kumar, S. Bashir and F. J. Chaloupka. 2012. "The Economics of Tobacco and Tobacco Taxation in Bangladesh." Report published by International Union Against Tuberculosis and Lung Disease, Paris, 53 pp.

Barkat, A., S. Zaman and S. Raihan. 2000. "Khas Land: A Study on Existing Law and Practice." Report published by Human Development Research Centre, Dhaka, 13 pp.

Baset, A. S. 2011. "Environmental Impact on Matamuhuri River by Tobacco and Shrimp Cultivation." Paper presented at the International workshop on impact of tobacco cultivation and policy advocacy for shifting to food and other agricultural crops, Dhaka, 28–30 March.

Brammer, H. 1997. *Agricultural Development Possibilities in Bangladesh*. Dhaka: The University Press Limited.

Buckles, D. 2008. "Reasons for Growing Tobacco in Daulatpur, Bangladesh." In *A Guide to Collaborative Inquiry and Social Engagement*, edited by J. Chevalier and D. Buckles, 153–60. New Delhi and Ottawa: SAGE Publications and IDRC.

Buckles et al., this volume.

Chevalier, J. and D. Buckles. 2013. *Participatory Action Research: Theory and Methods for Engaged Inquiry.* UK: Routledge.

Choudhury, P. 2010. "The British American Tobacco Company." Q3 performance update issued by BRAC, Dhaka. Online: http://www.bracepl.com/brokerage/research/1297917272BATBC_Jan11.pdf (accessed 20 December 2013).

Efroymson D. and S. Ahmed. 2001. "Hungry for Tobacco: An Analysis of the Economic Impact of Tobacco Consumption on the Poor in Bangladesh." *Tobacco Control* 10(3): 212–17.

Eizenberg, Hanan, Radi Aly and Yafit Cohen. 2012. "Technologies for Smart Chemical Control of Broomrape (*Orobanche* spp. and *Phelipanche* spp)." *Weed Science* 60(2): 316–23.

FAO (Food and Agriculture Organization of the United Nations). 2003. *Issues in the Global Tobacco Economy: Selected Case Studies.* Rome: FAO's Commodities and Trade Division, Raw Materials, Tropical and Horticultural Products Service.

———. 2012. *Food and Agriculture Organization (FAO) Statistical Yearbook 2012.* Rome: FAO.

Haque, Tito R. 2010. "UBINIG Research on the Matamuhuri River System." [in Bangla] Report published by UBINIG, Dhaka.

Hoque, M. E. 2001. "Crop Diversification in Bangladesh." In *Crop Diversification in the Asia-Pacific Region.* Bangkok: FAO Regional Office for Asia and the Pacific. Online: http://www.fao.org/docrep/003/x6906e/x6906e04.htm#crop%20diversification%20in%20bangladesh%20m.%20enamul%20hoque* (accessed 5 January 2013).

Kalam Azad, Abul. 2013. "Govt Gives in to Tobacco Industry." *The Daily Star* [newspaper], 5 March. Online: http://www.thedailystar.net/beta2/news/govt-gives-in-to-tobacco-industry/ (accessed 20 December 2013).

Karim, Sayed Badrul. 2013. "Monthly Media Monitoring Report – March 2013." Unpublished report by PROGGA Knowledge for Progress, Dhaka.

Lecours, this volume.

Lins, R., J. B. Colquhoun, C. M. Cole and C. A. Mallory-Smith. 2005. "Postemergence Small Broomrape (*Orobanche minor*) Control in Red Clover." *Weed Technology* 19(2): 411–15.

Maniruzzaman, F. M., K. R. Haque, M. A. Sattar, M. A. Wadud Mian and S. K. Paul. 2011. *Agricultural Research in Bangladesh in the 20th Century: Crops, Forestry, Livestock, Fisheries.* Dhaka: Bangladesh Academy of Agriculture, Bangladesh Agricultural Research Council.

Mishra, U. and S. K. Hossain. 2005. "Current Food Security and Challenges: Achieving 2015 MDG Hunger Milepost." Paper presented at the national workshop on Food Security in Bangladesh, Dhaka, Bangladesh, 19 October.

MOHFW (Ministry of Health and Family Welfare). 2009. "Global Adult Tobacco Survey – Bangladesh Report." Report published by Ministry of Health and Family Welfare, Government of the People's Republic of Bangladesh, Dhaka, 230 pp.

Pain, A., I. Hancock, S. Eden-Green and B. Everett. 2012. "Research and Evidence Collection on Issues Related to Articles 17 and 18 of the Framework Convention on Tobacco Control." Report published by DD International for British American Tobacco. Online: http://ddinternational.org.uk/viewProject.php?project=21 (accessed 7 December 2012).

Rahman, M. M. and S. I. Khan. 2005. "Food Security in Bangladesh: Food Availability." In *Food Security in Bangladesh: Papers Presented in the National Workshop.* Dhaka: Government of Bangladesh and World Food Program.

Rahman, S. 2010. "The Thriving Tobacco Industry." *Probe Newsmagazine* 9(39), published 18–24 March. Online: http://www.probenewsmagazine.com/index. php?index=2&contentId=4337.

Rahman, T. 2011. "Does Tobacco Cultivation Threaten Food Security in Bangladesh?" Paper presented at the International workshop on Impact of Tobacco Cultivation and Policy Advocacy for Shifting to Food and other Agricultural Crops, Dhaka, 28–30 March.

Roy, A., D. Efroymson, L. Jones, S. Ahmed, I. Arafat, R. Sarker and S. Fitzgerald. 2012. "Gainfully Employed? An Inquiry onto Bidi-Dependent Livelihoods in Bangladesh." *Tobacco Control* 21(3): 313–17.

Sobhan, R. 1995. *Experiences with Economic Reform: A Review of Bangladesh's Development.* Dhaka: Centre for Policy Dialogue and University Press Limited.

UBINIG. 2007. "Economic Performance of Tobacco and Alternative Rabi Crops." Technical report published by UBINIG, Dhaka.

_____. 2008. "Evaluation of Rabi Crops (Combination) Cultivation against Tobacco Production." Technical Report published by UBINIG, Dhaka.

_____. 2009. "Technical Report on Firewood Consumption in Tobacco Kilns" [in Bangla]. Technical report published by UBINIG, Dhaka.

_____. 2011. "Technical Report on Alternatives to Tobacco." Technical report published by UBINIG, Dhaka.

Uddin Bhuyan, Md. 2012. "Government Takes Initiative to Curtail Tobacco Production." *New Age.* 15 December. Online: http://newagebd.com/detail.php?date=2012-12-15&nid=33483#.UqYcm41inWI (accessed 15 December 2012).

Uddin Molla, Md. 2011. "Economics of Producing Multiple Rabi Crops." Paper presented at the International Workshop on Impact of Tobacco Cultivation and Policy Advocacy for Shifting to Food and Other Agricultural Crops, Dhaka University, Dhaka, Bangladesh, 28–30 March.

WHO. 2007. "Impact of Tobacco-Related Illness in Bangladesh." Report published by World Health Organization, South East Asia Region, New Delhi, 76 pp.

Chapter 6

SUBSTITUTING BAMBOO FOR TOBACCO IN SOUTH NYANZA REGION, KENYA

Jacob K. Kibwage, Godfrey W. Netondo and Peter O. Magati

Features of the Tobacco Industry in Kenya

As in many other African countries, tobacco in Kenya began as a product of British colonialism. In 1907, British American Tobacco (BAT) established a base in Mombasa, Kenya from which to market and distribute tobacco products in what are now Kenya, Uganda, Tanzania and the Democratic Republic of Congo. The establishment of a market for cigarettes in East Africa prompted the opening of a cigarette factory in Jinja, Uganda. By 1948, it was the largest in the region. After a series of expansions, including the purchase of the East African Tobacco Company in Tanzania, BAT opened a cigarette factory in Nairobi in 1957. Other companies eventually set up their own manufacturing centers. Kenya is currently the cigarette manufacturing and distribution hub for 17 African countries (Wanyonyi and Kimosop nd.; Kweyuh 1994; Patel et al. 2007).

To ensure a steady supply of raw tobacco leaf, BAT and other manufacturers created a system of contract buying with farmers. This involved the creation of tobacco markets, the supply of inputs and access to technical training and credit. These arrangements were modeled on similar systems in place for tea and sugarcane production and stimulated tobacco growing throughout the region. The industry that manufactures tobacco products in Kenya currently draws on tobacco production in Kenya, Uganda, Tanzania and Malawi. Despite international efforts to control tobacco use and production in Africa, the industry is also expanding into new areas such as South Sudan where tobacco production is set to begin on a large scale.

Kenyan tobacco growing today follows on more than four decades of commercial cultivation in the South Nyanza Region (former Nyanza Province) on the shores of Lake Victoria. Until recently, tobacco production in Kenya was concentrated there (Figure 6.1). Over the last decade it has expanded into parts of the Western Region and Eastern Region with high agricultural potential. By 2011, the South Nyanza Region, Eastern Region and Western Region had 31.4 percent, 25.0 percent and 43.6 percent of the total number of tobacco farmers in the country, respectively. At present, about 55,132 farmers are involved in tobacco production in Kenya (Table 6.1), an increase of 57 percent since 2006 (Patel et al. 2007). The country's gross production (in terms of earnings) of tobacco has also increased dramatically between 1981 and 2010 (Figure 6.2).

The rapid growth of Kenya's tobacco production and tobacco manufacturing exports is due to promotional campaigns in the Western and Eastern regions, investment in new cigarette factories and the emergence of new tobacco companies on the national scene. While BAT continues to be the

Figure 6.1. Tobacco-farming regions in Kenya

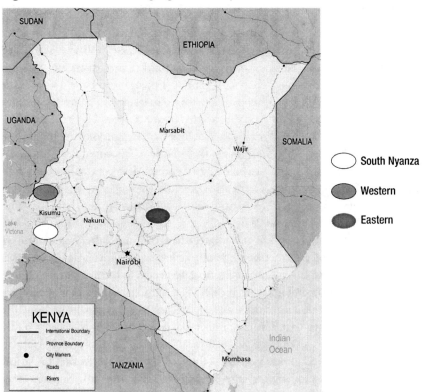

Table 6.1. Number of tobacco farmers in Kenya, 2011

Region	No. of contracted farmers	No. of independent farmers	Total farmers producing tobacco
South Nyanza	10,203	7,131	17,334 (31.4%)
Western	13,405	10,629	24,034 (43.6%)
Eastern	4,188	9,576	13,764 (25.0%)
Total	27,796 (50.4%)	27,336 (49.6%)	55,132 (100.0%)

Figure 6.2. Gross export of tobacco and tobacco products (KES millions), 1981–2010

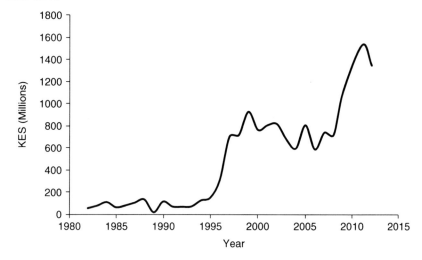

Source: Republic of Kenya 2012.

largest player, two other companies – Mastermind Tobacco and Alliance One Tobacco – have expanded their operations and developed direct relationships with tobacco farmers. Alliance One Tobacco has taken over the former BAT contracted farmers in Nyanza region under confidential transfer arrangements between the two companies. Farmers were not consulted during this process. Four new companies have focused on the manufacture and distribution of cigarettes. Cut Tobacco, Ozzbeco, BAT Equatorial Africa Area and McCroft Tobacco Holdings have taken a share of the tobacco products market, which has stimulated new production and export. Some of these companies are new to tobacco (Ozzbeco is also a beer maker) while others are subsidiaries of established tobacco companies (BAT Equatorial Africa Area).

Various national policy reforms and international market dynamics have also helped to create the conditions for the expansion of the tobacco industry in Kenya. Structural adjustment policies of the 1980s dramatically reduced public investment in agricultural research, extension and market supports, which provoked a generalized weakening of the agricultural economy in Kenya (Republic of Kenya 2010). Major shocks to agriculture followed in the early 1990s when the cotton, pyrethrum (chrysanthemum) and sisal industries in Kenya suffered a collapse in prices and supply chains due to stiff competition from imports. These crops were critical to both industrial and small-scale agriculture, leaving farmers with few options for generating cash income. During the same timeframe, tea, coffee, maize, wheat, rice, flowers and sugar cane suffered greatly from structural adjustment policies such as the elimination of fertilizer subsidies, deregulation of input and commodity prices and the closure of state-owned marketing facilities. Cultivated area per household of many of these crops fell dramatically even as domestic demand remained stable or grew. The crisis in fishing on Lake Victoria, provoked by the invasion of water hyacinth, also undermined livelihood strategies in some districts of the South Nyanza Region (Muyanga and Jayne 2006). Tobacco production, with its integrated system of input supply, assured markets and corporate organization filled the gap left by a narrowing of cash crop options for farmers, the generalized crisis in traditional livelihoods and fragmentation of traditional community organization.

In Suba, Kuria, Homa and Migori districts in the southern part of the former Nyanza Province, for example, tobacco farmers said that the establishment of Mastermind Tobacco Company in the early 1990s was a key event in the local evolution of tobacco production. A timeline analysis of tobacco production by a group of tobacco farmers interviewed by the authors showed as well that the crisis with other cash crops and the absence of local organization prompted many farmers to abandon maize, fish, livestock and groundnuts in favor of tobacco production.

Since the period of intensive structural adjustment in the 1990s, the Government of Kenya has made only modest investments in agriculture, even though the agricultural sector is recognized as the means of livelihood for most citizens and could potentially drive sustainable economic development. Policy documents such as the Strategy for Revitalization of Agriculture 2004–2014 (Republic of Kenya 2004) and its successors, the Agricultural Sector Development Strategy 2010–2020 (Republic of Kenya 2010) and the Kenya Vision 2030 Strategy (Republic of Kenya 2007) all call for significant growth in the agricultural sector. Little, however, has actually been done to create an enabling environment for broad-based agricultural diversification and growth. This continues to leave farmers in established tobacco-growing areas such as

Nyanza region and areas of expansion such as the agricultural lands of the Eastern and Western regions with few options and little scope to contribute in new ways to the development of the nation. Even traditional food crops like cassava, millet and sweet potatoes that are important in periods of drought and famine have been largely displaced from the tobacco-growing regions. Livestock has also drastically declined as grazing areas are converted to tobacco farms.

The agricultural policy environment is further complicated by a conflict of interest between different branches of government. On the one hand, the National Social Security Fund, an arm of the Government of Kenya, owns 20 percent of BAT operations in Kenya and helps to appoint members to the company's Management Board. Government coffers receive dividends on its shares in the company, in addition to taxing revenue on company profits. The total annual production for all tobacco companies in Kenya is valued at over one billion shillings (USD 13.8 million), although this is less than four percent of export earnings and is declining as other sectors of the economy grow (Patel et al. 2007).

On the other hand, the Ministry of Health is trying to control tobacco consumption and the Ministry of Agriculture has formally committed itself to discouraging tobacco production. The government's tobacco-control campaigns began in 1992 (Figure 6.3). In 2001, the Ministry of Health established the national Tobacco Free Initiative Committee to coordinate tobacco-control activities in the country, especially the World No Tobacco Day (WNTD). The Government of Kenya helped to negotiate the Framework Convention on Tobacco Control (FCTC) and ratified the treaty in 2004. In 2007, it passed a comprehensive Tobacco Control Act to control the production, manufacture, sale, labeling, advertising, promotion and sponsorship of tobacco products. The Act also established a Tobacco Control Board in 2008 to provide technical advice to the Ministry of Health on tobacco control and created a National Tobacco Control Plan in 2010. Despite these commitments to control tobacco use and regulate production, the national government's ongoing ownership in a share of the industry creates a profound ambivalence within government

Figure 6.3. Kenya tobacco-control timeline, 1992–2010

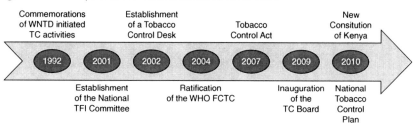

Source: WHO (2012).

towards the active development of alternatives to tobacco and makes the government complicit in the industry's current expansion within the country (Wanyonyi and Kimosop nd.).

Farmer Experiences with Tobacco Production in the South Nyanza Region

Research in Kenya raises serious concerns about the social, economic, health and environmental impacts of tobacco farming (Patel et al. 2007; Ochola and Kosura 2007; Abila 2006; Kweyuh 1994). The growing debate in Kenya regarding the role of tobacco farming and cigarette manufacturing in the country prompted the development of an action-research project focused on creating an alternative to tobacco production for farmers in the South Nyanza Region (former Nyanza Province). With the financial assistance of the International Development Research Centre (IDRC), a research team centered at the South Eastern Kenya University began to experiment with bamboo cultivation and to engage farmers in assessing its potential as a substitute for tobacco. This chapter presents results from research with several hundred farmers in the South Nyanza Region between 2005 and 2013. It also presents lessons from this research on the conditions that need to be in place for substitute crops to become economically viable and sustainable alternatives to tobacco.

Work on developing a substitute for tobacco production began in 2005 with a detailed and participatory assessment of farmer experiences with tobacco production in the Migori, Kuria, Homa Bay and Suba districts (the first two are located in Migori County and last two in Homa Bay County) near Lake Victoria. This is where tobacco production in Kenya was initially concentrated and where it remains a cash crop of importance. Resource mapping exercises, seasonal calendars and timelines of major events in the history of agricultural development in certain communities helped the research team formulate a number of working hypotheses regarding the local features and dynamics of the tobacco industry. These were examined further through a formal survey conducted in 2007[1] and through field experiments with bamboo. The team also undertook various studies on markets for bamboo products and organized farmers into producer cooperatives, whose activities are described below.

---

1   The survey, which involved 440 smallholder households, followed standard recommendations regarding design and implementation of a multi-stage and stratified random sampling of households in four districts of southern Nyanza region (Migori, Suba, Kuria and Homa Bay). One administrative location with the highest concentration of tobacco farmers was selected from each district.

The research shows that most farmers were drawn into tobacco production by the offer by tobacco companies of crop inputs and technical assistance, along with assurances that a ready buyer existed for their product (Table 6.2). The inputs, initially supplied at no or very low cost, included tobacco seed for flue- and fire-cured varieties of interest to the companies, fungicides, pesticides, fertilizers and processing materials such as cotton twine and bags. In subsequent years, the companies continued to supply these inputs on credit and deducted the cost each year from the sale of the finished product. Costs such as firewood, labor, land rental and depreciation of land due to intensive monoculture were left to the farmers to manage (Abila 2006).

Table 6.2. Reasons for starting tobacco farming, South Nyanza Region

Main reason for starting tobacco farming	Response (%) n =285
Anticipated a ready market	23.0
Tobacco company provided inputs	20.8
Promotion by agricultural officers	19.8
Influenced by other tobacco farmers	18.9
Inherited from ancestors	8.2
Availability of land	6.2
No other cash crop at the time	3.1
Total	100.0

The price farmers receive for their tobacco leaf product depends on an evaluation of its quality by company field technicians. Under-grading at the time of purchase is a common complaint of tobacco farmers. As farmers are not organized, no independent assessment is done. The Nyanza, Eastern and Western Tobacco Farmers Association (NEWTFA), the only organization claiming to represent tobacco farmers in Kenya, was founded and is funded by the tobacco companies, presumably as a measure to pre-empt independent organizing (Abila 2006). In addition to controlling the selection and price of inputs and products, technical advisors from the tobacco companies also define the actual tobacco production process, from seeding time to harvesting and sale (Figure 6.4).

The long duration of the production cycle also has an impact on farmers' independence. Flue-cured tobacco, the most widely grown tobacco in the study area, occupies the land and labor of farmers and their families for 10 months of the year, directly displacing most other crops and leaving farmers with little time to engage in other activities. Most farmers (90 percent of those

Figure 6.4. Cycle of tobacco farming in Kenya

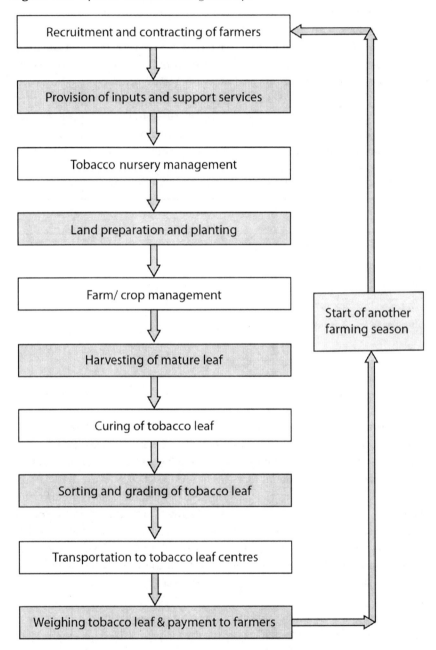

surveyed) have also invested in building their own curing barns. This represents an upfront investment that later creates a barrier to shifting out of tobacco production. As a result of these various technical and social arrangements, tobacco farmers are more specialized in and tightly bound to the crop than to other crops and other sources of livelihood. The survey data shows that 82 percent of tobacco farmers are engaged in farming only (not other livelihood activities). This is a higher percentage than within the general farming population, which tends to rely on non-agricultural activities to supplement their livelihoods.

Tobacco farmers also use their land more intensively and in a more specialized manner than the general farming population. The data shows that they allocate significantly less land to food crops, woodlots, Napier grass for livestock and fallow compared to other kinds of farmers (Kibwage et al. 2009). This high level of specialization increases tobacco farmers' exposure to the many complications associated with intensive and long-term monocultures, such as crop diseases, loss of soil organic matter and soil erosion. At the same time, placing all hopes on a single crop increases the risk of complete crop failure due to unexpected rainfall patterns or other unpredictable weather events, such as hail, excessive heat, etc.

The gender dimension of tobacco farming in Kenya is also notable. Our survey data shows that households headed by an adult female are less prominent among tobacco farmers as compared to the general farming population. Of 210 tobacco-growing households interviewed in 2007, only 14 percent were female-headed, compared to almost 33 percent among all farmers. This suggests that tobacco farming is more frequently pursued by men than by women farmers. Male tobacco farmers also tend to have a slightly higher rate of polygamy and larger families than the general farming population. This may be a way to create a home-based labor force that can be mobilized to meet the intensive and seasonal demand for labor associated with the planting and curing stages of tobacco growing.

Although the total farm area under tobacco cultivation in South Nyanza Region has been increasing steadily, farmers move in and out of tobacco farming for a variety of reasons. Active tobacco farmers surveyed (Table 6.3) expressed their dissatisfaction with tobacco prices and with the tobacco grading system, which they felt should be monitored by the government or a reliable farmer-based organization. They also commented on the tedious nature of the work, various occupational health problems and the risk of major accidents on the farm due to fire. The labor intensive nature of tobacco production and its many negative health impacts (smoke inhalation, eye diseases) were highlighted by active tobacco farmers (Table 6.3) and by farmers that had either abandoned tobacco production or decided not to start tobacco growing in the first place (Table 6.4).

Photograph 6.1. The two wives of one husband sorting tobacco leaves into graded piles after curing, Homa Bay County, Kenya

Photo credit: J. K. Kibwage.

Table 6.3. Curing and marketing problems facing active tobacco farmers, South Nyanza Region

Tobacco curing problems	% Response N = 194	Tobacco marketing problems	% Response N = 192
Tedious work	23.3	Low prices	25.3
Smoke inhalation	23.1	Poor grading	22.2
Too labor intensive	19.5	Delayed payments	14.0
Eye problems	18.1	Transportation problems	13.1
Barns catch fire	16.0	Company delays in buying tobacco	9.9
		Theft of tobacco bales	7.9
		Inadequate extension services	7.6
Total	100.0	Total	100.0

Objections to smoking tobacco among Christian religious leaders and their followers also plays a role in convincing some farmers to abandon tobacco (11 percent of respondents from this category) and other farmers to avoid tobacco growing in the first place (33.7 percent of respondents from this category).

Table 6.4. Reasons for not participating in tobacco production, South Nyanza Region

Main reasons for abandoning tobacco cultivation	% response N = 63	Main reasons for not growing tobacco at all	% response N = 107
Too labor intensive	31.4	Religion	33.7
Health-related issues	27.1	Health-related issues	29.1
Low returns	26.3	Too labor intensive	27.9
Religion	11.0	Scarcity of land	5.2
Scarcity of land	4.2	Lack of seeds and other farm inputs	4.1
Total	100.0	Total	100.0

Table 6.5 shows that tobacco production has greatly affected food production in the region, particularly maize. Some 50 percent of the farmers surveyed replaced their maize crops with tobacco. The remainder replaced other food crops such as beans, cassava, sweet potatoes, etc. According to farmers, chemicals used on tobacco farms have also had a negative effect on their ability to grow important food crops, such as vegetables. Furthermore, they noted that tobacco extracts nutrients from the soil at a very high rate compared to other crops and leaves no organic matter to be returned to the soil. In an effort to cope with food insecurity, tobacco farmers tend to buy or borrow food from their relatives or seek relief food from the government or aid agencies.

Farmers surveyed also questioned the economic returns from tobacco. The survey found that the annual net income of a non-tobacco farmer is typically higher than that of a tobacco farmer, with an average annual difference of USD 198. This is a significant difference in living standards at the local level.

The effects of income differentials can be seen at the household level when we compare the two groups. For example, housing among tobacco farmers is generally poorer, with a significantly larger proportion of tobacco farmers living in temporary homes with mud walls and roofs made of iron sheeting or thatch, as compared to non-tobacco farmers. The ownership of livestock, a key indicator of wealth in Kenyan society, is virtually the same in both populations, as is ownership of physical assets such as motorcycles, televisions, radios, etc. Differences in other expenditures are more evident, however.

Table 6.5. Impact of tobacco farming on other crops, South Nyanza Region

Main crop replaced by tobacco	% of respondents
Maize	50.4
Beans	9.4
Cassava	8.5
Sweet potatoes	6.8
Groundnuts	6.0
Tomatoes	6.0
Sorghum	5.1
Kales (*Sukuma wiki*)	3.4
Pineapples, bananas, finger millet (combined total)	4.4
Total	100.0

The survey data shows that tobacco farmers spend more income (USD 35 more) on average per year on medical and healthcare services than non-tobacco farmers. This suggests that tobacco households, on average, are more prone to illnesses requiring medical assistance. By contrast, non-tobacco households on average spend more of their income on education as compared to tobacco farmers. This suggests that in the longer term, farm families engaged in tobacco farming are less able to improve their situation through education for their children. This in turn puts them and their families at a disadvantage in the broader labor market. Even though farmers presumably engage in tobacco farming to improve their standard of living, tobacco farming seems to have contributed little to their livelihoods in terms of social status, asset ownership and intergenerational socioeconomic development.

The environmental impact of tobacco growing is also a concern in Kenya due to the demand for firewood to fuel the curing of tobacco leaves and the high rate of deforestation in tobacco-farming areas. The type of tobacco grown in Kenya is cured using wood fuel, which necessitates the felling of trees. While the tobacco industry supports reforestation with eucalyptus (a fast-growing exotic tree species with many known environmental problems) it also discourages farmers from using eucalyptus to cure tobacco because the smoke gives an undesirable smell to the dried leaves. Consequently, indigenous tree species take the worst of tobacco pressure on forests, leading to the loss of forest biodiversity, soil degradation and the long-term decline of water catchment areas (WHO 2012). In 2011, an environmental audit of tobacco companies resulted in a very poor grade because of their lack of compliance with the environmental standards established by the Kenya National Environment Management Authority (NEMA) and other organizations.

Developing Bamboo as an Alternative Cash Crop

Research in Kenya on economic alternatives to tobacco growing points to the need to broaden the range of crops available to farmers (Patel et al. 2007). To this end, the research team launched a research process with farmers, village elders and local Ministry of Agriculture officials in four districts in South Nyanza Region where tobacco farming is prominent. It focused on the cultivation of bamboo and comparisons with tobacco farming over a period of three years, from 2006 to 2009. These activities involved 240 farmers over the period.

The research embraced participatory and multi-stakeholder approaches to engaging with farmers and the communities where they live. In the heart of Kenya's traditional tobacco-growing region, the work built on positive grassroots political will to control tobacco. It involved local leaders from churches and non-governmental organizations, as well as democratically elected councilors and Members of Parliament. Care was taken to plan and schedule activities based on a thorough understanding of the local farming calendar and an assessment of farmers' readiness to participate. Special efforts were made to ensure that poor and women farmers with small parcels of land could participate in the project. Farmer organization was also emphasized, leading to the development of producer cooperatives around which production and marketing were organized.

Field trials

The research started with bamboo field trials. Bamboo was selected for experimentation because it has multiple economic uses and was reportedly well-suited to the soil and climate found in the region (Kenya Forestry Research Institute 2008; Kigomo 1988; Ongugo et al. 2000). The crop has the potential to yield 20 to 40 tons of poles per hectare each year with a growth rate three times faster than eucalyptus, a common forestry species with commercial value (Kibwage et al. 2008). Bamboo matures in three to four years and can sustain continuous harvesting for decades with very little investment after it becomes established (Karina 1998). Due to its light weight, high elasticity and great resistance to breakage, bamboo is suited to a variety of purposes. It can be used in the bio-fuel industry and in the production of pulp and paper, handicrafts and household goods. Bamboo leaves make good animal fodder and the shoots are edible to humans. The environmental benefits of bamboo as an agroforestry crop are many, including the conservation of soil and water, habitat protection, rehabilitation and stabilization of gullies and riverbeds and recycling and filtration of domestic and industrial wastewater.

Willingness to provide land for experiments with bamboo was strong among both tobacco and non-tobacco farmers in the area. Establishing experiments with both types of farmers helped the project avoid the perception that bamboo farming was only for tobacco farmers and made it possible to engage with a range of poor smallholders and women farmers. The project team and officials from the Kenyan Ministry of Social Services worked to establish and register with the government four community-based Bamboo Farmers Groups to undertake the experiments – one in each district. This enabled a systematic approach to technical training of farmers in bamboo cultivation and was a way to build confidence among farmers in their technical skills and group management skills. These groups were later formalized as Bamboo Farmer Cooperatives, thereby giving a unique identity to the farmer groups and creating a longer-term vision of themselves as bamboo producers.

To confirm the technical viability of the crop in the South Nyanza Region some 120 trial farms were established in 2006 using 2,420 bamboo seedlings of two different species, *B. vulgaris* and *D. giganteus*. These sympodial (lateral growth) types of bamboo were selected because they are smooth and highly resistant to wear, giving them a higher commercial value than monopodial types. The two species propagate as a clump, sending up shoots from around the base and gradually spreading across the ground. This makes the plant compatible with other crops that farmers can plant in-between the clumps, at least until a solid stand is established.

Growth performance studies over time showed high rates of seedling survival on a wide range of soil types and climatic conditions, including those where tobacco is grown. The average survival rate for *B. vulgaris* was 94 percent and for *D. giganteus* 70 percent. The number of culms (poles), culm height and culm diameter monitored over three years confirmed that both species could be harvested on a regular basis without putting the survival of the plantation at risk (Kibwage et al. 2008). Harvesting rates were as expected, averaging 2,295 culms per year from a farm with 200 clumps on one hectare of land (Magati et al. 2012).

While all of this was promising, the time to initial harvest of bamboo (three years) was obviously a concern to farmers. To manage this concern, the field trials introduced intercrops into the bamboo plantations, including kale and legumes such as beans and cowpeas. These crops produced normally during the first year, providing farmers with an economic benefit from their land as the bamboo crop grew. After 18 months, however, the bamboo clumps shaded the ground to such a degree that understory crops would not grow. Since none of the farmers had dedicated all of their cropland to the bamboo experiment, they managed income and cash flow by taking on other activities during this critical period. For some farmers, this included tobacco production on other lands.

Photograph 6.2. Bamboo inter-cropped with vegetables (various kales) in year one of bamboo growth to generate farm income and manage cash flow requirements while the bamboo develops, Migori County, Kenya

Photo credit: J. K. Kibwage.

Despite the success with growing bamboo, marketing of bamboo products proved to be a challenge. While Kenya has about 150,000 hectares of bamboo forests, mostly in Central, Rift Valley and Western regions of the country, a 1986 presidential ban on bamboo harvesting from government forests meant that very little Kenyan bamboo was entering the Kenyan market. When this experiment was launched, no farmers were actively growing bamboo as a commercial crop. Most bamboo available in Kenya was in the form of transformed products imported from China, India and Thailand (Kibwage et al. 2008). This included toothpicks, baskets, bowls, tablemats, trays, skewers, flower vases and edible shoots, among others. A few local manufacturers supplied baskets made locally from bamboo, although only in small quantities. Small quantities of bamboo in the form of dried poles also made it into the market as scaffolding through illegal extraction from government forests. However, the Kenya Government in late 2013 officially lifted this ban to enable the bamboo sector to expand, create jobs and improve both rural and urban livelihoods, a policy shift inspired in part by the success of the project experiments.

Feasibility studies conducted by the research team suggested that bamboo could be marketed to the housing and construction industry for scaffolding and as a construction material. Eucalyptus tree poles, the main material used for scaffolding in Kenya, is expensive and of inferior quality compared to bamboo (Ongugo et al. 2000). The studies determined that simple drying procedures using existing tobacco kilns familiar to tobacco farmers were sufficient to create a market-ready product (a dried pole). Other promising markets for bamboo products identified through the studies included bamboo furniture (made from dried poles), bamboo weavings and handicrafts (made from raw poles) and bamboo seedlings (for sale in nurseries to new producers). All of these products could be created through small-scale cottage industries or by farming households themselves, drawing on existing skills with crafts and plant materials.

The research also showed that the base value of raw bamboo poles increased dramatically when transformed into these higher-value products.

Photograph 6.3. Women making bamboo mats and baskets to enhance their income, Homa Bay County, Kenya

Photo credit: J. K. Kibwage.

The production of bamboo seedlings increased the base value by 50 percent, drying poles for construction by 100 percent, furniture construction by 200 percent and handicrafts by 400 percent. These gains held equally for the three major urban markets in Kenya: Nairobi, Mombasa and Kisumu (Kibwage et al. 2013). To support further development of value-added products from bamboo, the Bamboo Farmers' Cooperatives set up by the project established bamboo processing and training facilities that use infrastructure previously used to process tobacco. Their goal is to work with members of the cooperatives to market bamboo products, especially scaffolding poles, furniture and handicrafts.

Comparisons between tobacco and bamboo

The research with farmer groups created the conditions for comparing the economic performance of tobacco with bamboo. Using crop performance data from the field trials, household survey data from 2007 and actual market prices for tobacco and bamboo poles for 2006 and 2007, the research team estimated the economic and financial benefits and costs of bamboo as a substitute for the tobacco crop (Magati et al. 2012). The comparison showed that the annual estimated income from bamboo farming is four to five times higher than for tobacco at farm-gate prices on the same land area. Labor costs are also lower for bamboo production, at 179 person-days in the first season on average, declining markedly in subsequent years. This compares to an average of 227 person-days per season required for tobacco farming. In most cases, the labor needs for bamboo production can be met from within the household, rather than relying on hired labor. Holding other factors constant, this leaves the household with 48 more person-days to diversify to other income-generating activities.

Estimates of the net value of the two crops showed rates of return more than 300 percent higher for bamboo farmers, at KES 663,272 per hectare compared to KES 155,445 per hectare for tobacco farmers (Magati et al. 2012). Even higher rates of return can be realized from the same level of bamboo production if the bamboo products are transformed into higher-value products like bamboo furniture, housing construction materials, assorted handicrafts or high-value charcoal. Table 6.6 summarizes the main contrasts between tobacco and bamboo emerging from the research.

The technical features of bamboo production – planting of seedlings in clumps on fields prepared for intercropping – is simple and straightforward for farmers to execute. Intercropping of bamboo with vegetables, legumes, peppers and other horticultural crops in the first year of the experiment gave farmers a source of income as they waited for the bamboo to mature in the third year.

Table 6.6. Contrasts between tobacco and bamboo

Tobacco	Bamboo
Has only one use (smoking and chewing) hazardous to human health.	Has over 2,000 documented uses worldwide, including industrial, construction, pharmaceutical, food, conservation, etc.
Annual yields associated with poor economic returns on labor (**KES** 6,000 per hectare).	Annual yields of about 20 to 40 tons per hectare in a well-managed plantation (KES 83,910 per hectare).
Consumes large amount of wood fuel.	Mature bamboo (three years and older) does not require any treatment before use. Immature bamboo (two to three years old) can be dried using bamboo residues only, in existing tobacco kilns.
Fertilizers and pesticides create soil and water pollution.	Purifies air and polluted water bodies. No fertilizers or chemicals needed at the farm level.
Extracts important plant nutrients from the soil, leaving it almost barren.	Supports bio-remediation and improves soil fertility thanks to decomposing leaves and sheaths.
High risk of crop loss due to natural calamities (hail, disease and fire outbreaks).	Few risks associated with natural calamities.
Labor intensive. Stimulates the use of child labor, especially for harvesting and curing.	Not labor intensive, after the first year. No child labor required during harvesting and processing.
Matures in about six to seven months but requires significant annual investments.	Matures in about three to four years but can be harvested continuously thereafter with little ongoing investment.
Prompts deforestation and soil erosion.	Good for soil stabilization and river bank protection.
No scope for on-farm or community-based transformation of products.	Creates community-based processing and transformation jobs.
Tobacco leaves take three to five days to process in a curing barn, under strict monitoring by the farmer.	Bamboo poles take about 20 minutes to be treated if done in the same tobacco curing barn under same temperatures. Does not require close monitoring by the farmer.

Photograph 6.4. Farmer harvesting bamboo from a former tobacco field

Photo credit: J. K. Kibwage.

Table 6.7. Farmers experimenting with bamboo who abandoned tobacco farming, 2006–2013

Cooperative name	Number of tobacco farmers experimenting with bamboo in 2006	Participating farmers continuing to grow tobacco in 2013	% change
Suba Bamboo Farmers' Cooperative Society Ltd	30	0	100.0
Homa-Bay Bamboo Farmers' Cooperative Society Ltd	18	4	77.8
Kuria Bamboo Farmers' Cooperative Society Ltd	58	27	53.4
Migori Bamboo Farmers' Cooperative Society Ltd	16	1	93.8
Total	122	32	73.8

Educational work and demonstrations showed farmers that bamboo has many potential uses. This reduced farmers' concerns that they would end up with a product they could not use. Positive farmer group dynamics helped enormously with farmer mobilization, instilling patience and discipline among farmers and engaging them actively in monitoring progress and responding to course adjustments in light of organizational problems encountered along the way.

The experience of growing bamboo was compelling for many farmers. More than half of the tobacco farmers participating in the field trials continued to dedicate sizable parts of their fields to bamboo production well after the research project ended. Most of the non-tobacco farmers involved in the trials also maintained or expanded their bamboo plantations and remained committed to the bamboo cooperatives. Importantly, almost three-quarters of the tobacco farmers involved in the experiments abandoned tobacco farming altogether, even though the industry continued to expand in the region (Table 6.7). The exception was Suba, a district where tobacco farming has deep roots and where bamboo did not grow well due to poor climatic conditions.

Conclusions

Tobacco production emerged in Kenya because cigarette manufacturers in need of a steady supply of raw material actively promoted the crop. The companies provided tobacco seed and other inputs and technical assistance so farmers could acquire the skills and knowledge to grow tobacco. These investments took advantage of periods of general decline in the agricultural sector in Kenya and weakening government investment in agricultural research and extension from the 1980s to the present. In the absence of any effort to organize the farming population, farmers remained fragmented and unable to apply pressure on government agencies to invest in alternatives or protect their interests.

Despite clear commitments from the Government of Kenya to control tobacco consumption and to develop alternatives for tobacco farmers, the industry has continued to grow rapidly. It has expanded into agriculturally rich parts of the country as well as neighboring countries, despite the evidence reported here showing that tobacco farmers have higher health costs, fewer children in school and similar or lower net income compared to non-tobacco farmers. The expansion presents a clear danger to the forests and major food-producing areas of Kenya and offers little of lasting value to farming communities and Kenyan society at large.

The development of alternatives to tobacco and efforts to encourage farmers to stay out of tobacco farming is likely to require investments similar to those the tobacco industry offered when promoting its crop. The research reported here suggests that the organization of farmers into cooperatives, combined with the offer of strategic inputs (in this case, adapted bamboo seedlings), training in production techniques and strategic marketing support achieved a high level of farmer commitment to bamboo production as an alternative to tobacco. Farmers' willingness to organize and shift at least some of their land out of tobacco production confirms the relevance and potential of bamboo from their point of view. The systematic comparison of tobacco and bamboo production also shows that the alternative is more profitable under these conditions.

Developing partnerships with public and private sector investment in sustainable bamboo production and the transformation of bamboo into higher-value products remain major challenges. While it was beyond the scope of a single research project to deal with all these issues, the study contributed to the development of a National Bamboo Industry Development Strategic Paper, under the authorship of the Ministry of Water, Environment and Natural Resources. The paper informed the government in a review that led to a lifting of the ban on bamboo harvesting from the government's national forests. This is expected to add impetus to the Kenyan bamboo sector in general. It will also justify additional research on production methods and bamboo varieties under varying soil and climatic conditions and stimulate markets for a host of bamboo products.

Other crops may also be viable alternatives with supports of this nature and building on lessons from the research reported here. While the government may not be able to immediately provide for other crops a service model as complete as is currently offered by tobacco companies, organized farmers can draw on existing agricultural development programs in their regions and demand services that will help them generate income and diversify their production systems.

References

Abila, R. A. 2006. "Tobacco in Kuria District, Status, Impacts and Policy Issues." Consultancy report for Action Aid, Kenya. 41 pp.

Karina, N. Q. 1998. "Ancient Grass, Future Natural Resource: The National Bamboo Project of Costa Rica: A Case Study of the Role of Bamboo in International Development." Issue 16 of an INBAR working paper. International Network for Bamboo and Rattan, Beijing, 58 pp.

Kenya Forestry Research Institute. 2008. *Status of Bamboo Resources Development in Kenya.* Nairobi: Kenya Forestry Research Institute.

Kibwage, J. K, A. J. Odondo and G. M. Momanyi. 2009. "Assessment of Livelihood Assets and Strategies among Tobacco and Non Tobacco Growing Households in South Nyanza Region, Kenya." *African Journal of Agricultural Research* 4(4): 294–304.

Kibwage, J. K, G. W. Netondo, A. J. Odondo, B. O. Oindo, G. M. Momanyi and F. Jinhe. 2008. "Growth Performance of Bamboo in Tobacco-Growing Regions in South Nyanza, Kenya." *African Journal of Agricultural Research* 3(19): 716–24.

Kibwage, J. K, G. W. Netondo, P. O. Magati, F. M. Mutiso, L. B. Marwa and C. M. Siocha. 2013. "Bamboo Production as an Alternative Crop and Livelihood Strategy for Tobacco Smallholder Farmers in South Nyanza, Kenya." Final Technical Report. Ottawa: International Development Research Centre (IDRC), Canada, 55 pp.

Kigomo, B. N. 1988. *Distribution, Cultivation and Research Status of Bamboo in Eastern Africa.* Nairobi: Kenya Forestry Research Institute.

Kweyuh, P. H. M. 1994. "Tobacco Expansion in Kenya: The Socio-Ecological losses." *Tobacco Control* 3(3): 248–51.

Magati, P. O., J. K. Kibwage, S. G. Omondi, G. Ruigu and W. Omwansa. 2012. "A Cost-Benefit Analysis of Substituting Bamboo for Tobacco: A Case Study of Smallholder Tobacco Farmers in South Nyanza, Kenya." *Science Journal of Agricultural Research & Management* ISSN: 2276-6375. Online: http://www.sjpub.org/sjar/abstract/sjarm-204.html (accessed 20 December 2013).

Muyanga, M. and T. S. Jayne. 2006. "Agricultural Extension in Kenya: Practice and Policy Lessons." Working Paper 26 published by the Tegemeo Institute of Agricultural Policy and Development, Egerton University, Nairobi, 35 pp.

Ochola, S. and W. Kosura. 2007. "Case Study on Tobacco Cultivation and Possible Alternative Crops in Kenya." Study presented at the first meeting of the Ad Hoc Study Group on Alternative Crops (established by the Conference of the Parties to the WHO Framework Convention on Tobacco Control), Brasilia, 27–28 February.

Ongugo, P. O, G. O. Sigu, J. G. Kariuki, A. M. Luvanda and B. N. Kigomo. 2000. "Production-to-Consumption Systems: A Case Study of the Bamboo Sector in Kenya." Working paper No. 27 published by INBAR's in conjunction with Kenya Forestry Research Institute, Nairobi, 53 pp.

Patel, P., J. Collin and A. B. Gilmore. 2007. "The Law Was Actually Drafted by Us but the Government Is to Be Congratulated on its Wise Actions: British American Tobacco and Public Policy in Kenya." *Tobacco Control* 16(1): 72.

Republic of Kenya. 1975–2012. *Statistical Abstracts of Kenya.* Nairobi: Government Printer.

———. 2004. *The Strategy for Revitalization of Agriculture 2004–2014.* Nairobi: Government Printer.

———. 2007. *Kenya Vision 2030 Strategy.* Nairobi: Government Printer.

———. 2010. *Agricultural Sector Development Strategy 2010–2020.* Nairobi: Government Printer.

Wanyonyi, E. and V. Kimosop. nd. "Monitoring Implementation of Tobacco Control in Kenya: Identifying Gaps and Opportunities." International Institute for Legislative Affairs. Online: http://ilakenya.org/wp-content/uploads/2013/01/monitoringbrief1.pdf (accessed 22 December 2013).

WHO (World Health Organization). 2012. *Joint National Capacity Assessment on the Implementation of Effective Tobacco Control Policies in Kenya.* Geneva: World Health Organization.

Chapter 7

DIVERSIFICATION STRATEGIES FOR TOBACCO FARMERS: LESSONS FROM BRAZIL

Guilherme Eidt Gonçalves de Almeida

Introduction

In 2008, a World Health Organization study on economically sustainable alternatives to tobacco growing (WHO 2008, paragraphs 27 and 32–34) concluded that simply replacing tobacco with another cash crop was unlikely to be sufficient to sustainably reduce the vulnerability of small tobacco farmers and improve their quality of life. They argued that the diversification of agricultural activities and the development of non-agricultural opportunities would be needed to improve the livelihoods of tobacco farmers and facilitate their transition out of tobacco farming.

This chapter examines Brazil's attempts to move in this direction. It starts with a review of the main features of tobacco production in Brazil, including its strong orientation towards export markets, and the challenges to livelihood diversification these features present. The perspective of tobacco farmers is then explored by delving into explanations of why they grow tobacco and an assessment of the economic gains and losses of tobacco farming compared to more diversified farming systems. These assessments form the background for an in-depth examination of Brazil's National Program for Diversification in Tobacco-Growing Areas and the reasons why it has not yet achieved its potential. The chapter argues that the program lacks policy coherence and in particular fails to recognize that livelihood diversification requires a territorial approach that goes beyond the boundaries of current tobacco-growing areas. Further development of the program will need to create broader political conditions for livelihood diversification while at the same time making use of existing policies for sustainable rural development.

Features of Brazilian Tobacco Production

Brazil is currently the second largest producer of tobacco and the world's largest tobacco leaf exporter (Eriksen et al. 2012). In 2011 the country exported 541,000 tons of leaf to 100 different countries, at a value of over USD 2.9 billion. More than 84 percent of its national production was exported.

The prominence of Brazil in international markets for tobacco leaf is a relatively recent phenomenon, and the result of the active search by international tobacco companies for new production centers in developing countries (World Bank 1999; FAO 2003). This growth was built, however, on a long history of tobacco farming in Brazil going back to the Colonial period. According to Prado (2000), commercial cultivation of tobacco in Brazil began during the seventeenth century, when it became the third most important export (after sugar and gold). Even then it represented a huge proportion of Brazil's foreign trade. Furtado (2000) notes that much of this production was intended for Africa, particularly in exchange for slaves used in Bahia (on the Atlantic coast). In 1815, due to England's pressure to end the slave trade, Brazilian tobacco shifted from markets in Africa to markets in Europe (Furtado 2000).

At the time, although some tobacco was cultivated throughout Brazil, its main location was around the city of Cachoeira in Bahia state. Some tobacco was also produced in the bay islands and in the coastal region south of the state of Rio de Janeiro. Two other areas of cultivation were coastal parts of São Paulo state and in the southern state of Minas Gerais. Tobacco growing was seen at that time as a more "advanced" crop compared to other examples of colonial agriculture (Prado 2000). The crop required skill and careful management (for transplantation, protection against excessive sunlight, repeated and periodic pruning and removal of caterpillars) and fitted well with the presence of abundant and cheap slave labor.

In 1824, with the arrival of the first German immigrants, tobacco cultivation increased in other parts of southern Brazil. The immigrants settled on small farms that made intensive use of family labor. Initially devoted to supplying local demand, tobacco gradually became a regional product for trade and export. This evolution coincided with industrialization in Santa Cruz do Sul, the main city in the Rio Pardo Valley, and the gradual integration of local agriculture into the national economy. Industrialization provided support to tobacco processing and various improvements in tobacco cultivation (Vargas et al. 1998). In the nineteenth and early twentieth centuries, the land ownership structure in this tobacco enclave was based on small properties where farmers had the knowledge and skilled labor to grow tobacco using a minimum of infrastructure for marketing. The companies that dealt with these farmers were mostly domestic.

Photograph 7.1. Farmer without protection equipment applying pesticide to control suckers, São Lourenço do Sul, Rio Grande do Sul

Photo credit: G. E. G. Almeida.

In the 1970s, a trade embargo imposed on Rhodesia (now Zimbabwe) provoked a decline in Rhodesian tobacco farms, which had been the largest suppliers of tobacco to the European market. In response, transnational tobacco companies moved into the Rio Pardo Valley (Vargas et al. 1998; Vargas and Campos 2005). This shift was also part of a broader process of concentration of capital and operational changes in the transnational tobacco corporations, prompted by the expansion of the global market for leaf tobacco and cigarettes. In Brazil, the companies introduced new technologies for tobacco growing and set up new infrastructure and administrative systems that allowed them to expand and displace national companies. The success of the investment depended on having the right environmental conditions for tobacco production, high quality tobacco leaf and relatively low production costs. The companies also benefited from conservative economic and political policies and a regulatory environment that favored the development of agricultural commerce on a large scale (Chonchol 1986). An oligopsony emerged, that is, a situation characterized by many sellers of tobacco leaf and a small number of tobacco buyers. The main companies were British American Tobacco (BAT)/Souza Cruz, Universal Leaf Tobacco, Alliance

One International, Philip Morris and Japan Tobacco International (Silveira and Dornelles 2010).

Today, Rio Pardo Valley generates 97 percent of the tobacco production in Brazil, with the breakdown within the region as follows: Paraná 17.4 percent; Santa Catarina 27.1 percent; and Rio Grande do Sul 52.5 percent (Bonato et al. 2010). Figure 7.1 shows the evolution of tobacco production in Brazil during the 20 years between 1990 and 2010. According to the Brazilian Institute of Geography and Statistics (IBGE), the rise in national production

Figure 7.1. Evolution of tobacco production in Brazil and four Southern states, 1990–2010 (tons)

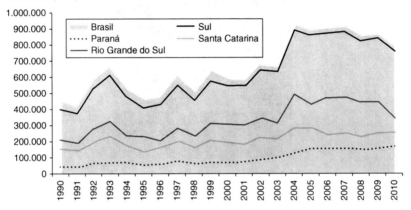

Source: IBGE (2010a). Online: http://www.sidra.ibge.gov.br (accessed 14 August 2012).

Figure 7.2. Evolution of Brazilian tobacco export by quantity (tons) and value (USD), 1999–2011

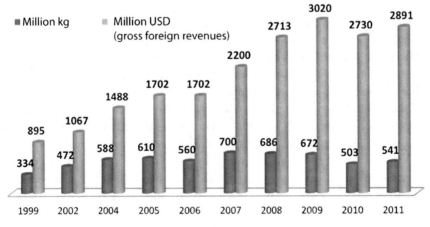

Source: Sinditabaco (2012). Online: www.sindifumo.com.br (accessed 30 June 2012).

was 372 percent, going from 182,915 tons of tobacco in 1975 to the highest level so far of 862,763 tons in 2005. Production increased by a staggering 85 percent between 2000 and 2005 alone.

Figure 7.2 shows the fluctuation of tobacco production as reflected in Brazilian exports between 1999 and 2011, while Table 7.1 compares exports in the three years between 2009 and 2011. Among the major importers of tobacco from Brazil, the European Union (EU) and Russia were highest, acquiring 40 percent of the volume exported. China and the United States were also major buyers of Brazilian tobacco.

The export of tobacco is directly influenced by exchange rates. When the Brazilian currency (the Real) depreciates against the U.S. dollar, the price paid to tobacco producers and the export price also drops. This makes Brazilian tobacco cheaper in dollar terms, helping the industry gain new markets and improve its competitiveness in foreign trade terms. The strong presence of Brazilian tobacco in the international market also reflects a national policy of exemption of tobacco leaf from the Tax on Circulation of Goods and Services (a tax on the export of primary products and semi-finished manufactured products).

While macro-economic conditions are key drivers of the growth of the tobacco industry in Brazil, micro-economic factors also have an influence. Tobacco farming in the Rio Pardo Valley, where much of national production is concentrated, is primarily undertaken by smallholders. Currently, about 35 percent of tobacco farmers in the Rio Pardo Valley own between 1 and 10 hectares of land, 25 percent own between 11 and 20 hectares of land and 25 percent are landless leaseholders. The remainder (15 percent) have larger properties. On average, these growers cultivate only 2.5 hectares of tobacco on their farms (AFUBRA 2012). For the tobacco companies operating in Brazil, this land tenure system carries no need for capital investments related to land acquisition. It also means that the companies do not need to hire workers to perform tasks that smallholder households can complete by using the unpaid labor available from within their families. This helps to reduce the costs of production and the price tobacco companies need to pay producers.

Access to financial markets by the tobacco companies was also important to the initial growth of the industry. In Brazil, the National Bank for Economic and Social Development (BNDES) actively financed tobacco industry investments in new plants, better processing capacity and the operational integration of different units within the same corporation. Not coincidentally, membership on the boards of transnational tobacco companies still includes representatives of financial institutions, pension funds and others from the financial services sector (Silveira and Dornelles 2010).

Investment allowed technological innovation to also spur expansion (Silveira and Dornelles 2010). In recent years companies introduced new production

Table 7.1. Value, quantity and price paid (by kilogram) for tobacco exported by Brazil, 2009–2011

Country	2009			2010			2011		
	Value (mil USD)	Weight (t)	USD/kg	Value (mil USD)	Weight (t)	USD/kg	Value (mil USD)	Weight (t)	USD/kg
Belgium	644,329	137,356	4.69	498,887	81,931	6.09	359,572	70,007	5.14
China	368,456	57,578	6.40	343,342	44,035	7.80	379,964	52,932	7.18
United States	308,093	66,407	4.64	242,113	52,845	4.58	276,760	58,645	4.72
Netherlands	161,183	37,579	4.29	198,232	39,214	5.06	199,019	31,702	6.28
Germany	175,798	35,157	5.00	187,333	28,862	6.49	168,666	28,454	5.93
Russia	116,306	39,088	2.98	119,374	30,278	3.94	190,542	41,390	4.60
Indonesia	106,053	16,950	6.26	82,704	11,929	6.93	134,859	22,442	6.01
Poland	57,889	15,514	3.73	79,836	16,198	4.93	100,660	16,774	6.00
Paraguay	63,368	21,853	2.90	63,244	15,795	4.00	67,700	17,512	3.87
Turkey	42,399	9,405	4.51	58,684	9,536	6.15	43,767	6,438	6.80
Total	3,046,032	674,731	4.51	2,762,246	505,620	5.46	2,935,187	545,603	5.38

Source: AGROSTAT 2012.

practices involving the use of seeds with genetic improvements, a specialized nursery for the cultivation of seedlings and the use of new compounds for fertilizers and pesticides. At the same time, the curing of tobacco leaves was improved by introducing new electronic instruments to control temperature and humidity and new packaging systems. These, in turn, have led to the use of new systems of hot air circulation, new types of furnaces and new energy sources, such as electric stoves. The effect of these changes has been to increase productivity from 1.92 Kg/ha in 2005 to 2.24 Kg/ha in 2012 (AFUBRA 2013).

Silveira and Dornelles (2010) have argued that transnational tobacco corporations make use of contract farming, known as the Integrated Production System (IPS), to enhance their bargaining power. In recent years the corporate strategy has been to *decrease* the number of farmers producing tobacco while increasing overall productivity. This process has led to a higher concentration of growers in certain rural communities, especially in the South of the country. In 2011 production increased while the number of growers decreased by 20,000.

Companies also have considerable influence over tobacco leaf prices determined within the Joint Commission, a formal entity in which representatives of industry and growers set prices and production conditions. The trading price is negotiated company by company, which in theory allows for competitive pricing between them. However, the industry uses the system of classification and grading of leaf to determine the price actually paid to growers at buying facilities. The cost of travelling to these facilities and transporting product is borne by farmers, dissuading some from participating in the grading process or taking back their product if they are not satisfied with the price. These transaction costs and the obligation to repay advances on the crop put the growers in a weak bargaining position vis-à-vis the buyers (Almeida, 2005). Price differentials in 2008–2009 help to illustrate the imbalance in power. Production costs that year were around USD 2.29/kg while the amount paid to farmers averaged only USD 2.67/kg, leaving a small profit margin. The export price for leaf tobacco that same year stood at USD 4.53/kg.

The above findings on the causes and factors influencing demand and prices for Brazilian tobacco leaf reinforce the argument made elsewhere in this book (Chaaban, this volume; Buckles et al., this volume) that domestic tobacco-control policies have little influence on these key farmer considerations. This fact is illustrated as well by Figure 7.3, showing an increase in leaf production of 34 percent between 2000 and 2010 accompanied by a drop of 33 percent in domestic cigarette consumption during the same period. Clearly, tobacco control in Brazil has not affected Brazilian tobacco farmers.

Figure 7.3. Evolution of tobacco production and cigarette consumption in Brazil, 2000–2010

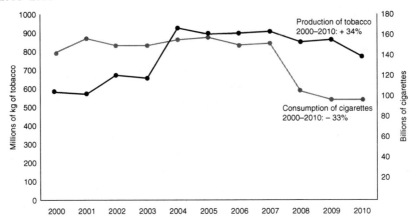

Source: Iglesias and Biz (2011).

Why Farmers Grow Tobacco

In 2012 some 165,170 households were involved in tobacco farming in Brazil. Some are specialized in tobacco farming and others are not. Vargas and Oliveira (2012) compared two groups of tobacco farmers: those who specialize mainly in tobacco production as their main source of income and those with more diversified production (including tobacco). According to their research, farmers specialized in tobacco farming in 2007–2008 had an average gross revenue from tobacco of BRL 39,616. This accounted for 96 percent of their total gross revenue that year. The production of corn ranked a distant second as a proportion of gross revenue (about one percent), while milk production ranked third (0.4 percent).

In the case of farmers who grow crops other than tobacco as their main source of income, the largest average gross revenue was obtained among farmers growing fruits and vegetables. As a group, these producers earned on average BRL 12,141 from this source in 2007–2008, representing about 40 percent of their total annual gross revenue. Tobacco production appeared as the second largest source of income for these farmers, accounting for 12.6 percent of their annual gross revenue (BRL 3,937). Milk ranked third, standing at 10 percent of annual gross revenue (Vargas and Oliveira 2012).

Table 7.2 shows that the productivity of tobacco farming, measured as gross value per hectare of land, is high compared to other crops. The analysis of net profit at the farm level paints a different picture, however. The average annual production cost for tobacco estimated by Vargas and Oliveira was

Table 7.2. Area harvested, yield, production value and economic yield per area (in thousands of BRLs) of the main crops grown in Brazil in 2000 and 2010

Product	Harvest (ha)		Production (tons)		Value of production (thousand BRL)		Productivity (thousand BRL/ha)
	2000	2010	2000	2010	2000	2010	2010
Soybeans	13,656,771	23,327,296	32, 820,826	68,756,343	8,658,735	37,380,845	1.6
Sugar cane	4,804,511	9,076,706	326,121,011	717,462,101	6,652,318	28,313,638	3.1
Milk					19,767,206	21,210,252	
Corn	11,890,376	12,683,415	32,321,000	55,394,801	6,037,136	15,186,463	1.2
Coffee	2,267,968	2,158,564	3,807,124	2,906,315	4,299,427	11,577,933	5.4
Manioc	1,709,315	1,787,467	23,044,190	24,524,318	2,585,287	6,896,070	3.8
Rice	3,664,804	2,722,459	11,134,588	11,235,986	2,586,649	6,242,880	2.2
Orange	856,422	775,881	106,651,289	18,101,708	1,262,673	6,021,746	7.2
Beans	4,332,545	3,423,646	3,056,289	3,158,905	1,658,867	4,938,454	1.4
Tobacco	310,462	449,629	579,727	787,617	1,022,024	4,508,061	10.0
Cotton	801,618	829,753	2,007,102	2,949,845	1,274,249	4,130,087	5.0

Source: IBGE (2010b).

Photograph 7.2. Family harvesting tobacco, São Lourenço do Sul, Rio Grande do Sul

Photo credit: G. E. G. Almeida.

BRL 23,582. This included both the variable costs (87 percent of total production costs) and the fixed costs of production (13 percent). Variable costs were defined as expenses incurred from permanent labor and agricultural inputs, agricultural operations with implements and others. These costs did not include the estimated cost of using family labor. Fixed costs mainly related to depreciation that underpins investment in the replacement of fixed assets used in agricultural production, in addition to spending to correct soil problems. Among diversified producers, the average annual production cost was BRL 11,211. Of this amount, 77 percent was dedicated to variable costs (BRL 8,589) and 23 percent to fixed costs (Vargas and Oliveira 2012). Their comparison with specialized tobacco farmers showed that diversified farmers had annual production costs that were, on average, about half of those incurred by their specialized counterparts.

The contrast is apparent. On the one hand, tobacco growing provides the highest gross revenue of any crop. On the other hand, production costs for specialized tobacco farmers are much higher than for more diversified farmers (even when family labor costs are treated equally). Taking into account both revenue and costs, Vargas and Oliveira (2012) concluded that the net financial gains obtained by specialized tobacco farmers (BRL 17,571) is 14 percent lower than for more diversified farmers (BRL 20,064).

Given these differences in net profits, why do farmers continue to grow tobacco when they could have higher net profits from other crops? The answer must take into account the unique features of the Integrated Production System (IPS) established by the tobacco companies. Under this system, tobacco farmers are contracted by companies to produce tobacco in specified quantities and in accordance with the technical instructions defined by the companies. The contract includes the supply of most inputs, technical assistance, loans brokered with banks, tobacco transportation from farms to buying houses and the purchase of all of the tobacco the farmers produce (Beling 2003). This is a complete technological and financial package, unavailable to farmers for any other agricultural product. It is not surprising then that tobacco farmers interviewed by Agostinetto et al. (2000) in the municipality of Pelotas (in Rio Grande do Sul) emphasized the security of income from tobacco compared with traditional food crops. They said that the fluctuation in demand and prices between successive harvests of other crops such as onions, potatoes and corn made these less attractive compared to tobacco.

The implications of the IPS for farmer livelihoods go beyond the simple calculation of costs and benefits. Carvalho (2005, 212) says that the farmer operating under the IPS, "knows his limits […] despite producing a salable commodity for the market." Specifically, they are acutely aware of the balance of power in the market and the risk of exclusion from the renewal of contracts and the manipulation of tobacco prices at the local level through the grading process. Generally, in periods when national supply of tobacco is high, the trend has been for companies to downgrade the product delivered by farmers, thus depreciating the value of tobacco production and providing farmers with a lower return than expected. When the national supply of tobacco is lower, the companies grade tobacco leaves more generously, resulting in higher payments to farmers (Silveira and Dornelles 2010). As demand for tobacco leaf on international markets is relatively constant, domestic supply ends up being decisive in determining the share of profits between companies and farmers.

Contractual arrangements and downgrading the classification of tobacco leaves also keep farmers indebted. In 2004–2005, when Brazil recorded tobacco production of more than 850,000 tons, the total amount of debt linked to tobacco marketing (the sum of the credit negotiated through tobacco companies) was estimated to be 48 percent of the tobacco growers' income (Buainain and Souza Filha 2009). These authors also estimated that between seven and 13 percent of tobacco farmers were indebted. The rates of indebtedness may be much higher, however. So far the tobacco companies have succeeded in suppressing access to the information needed to definitively assess the degree of indebtedness in the industry.[1] What is known anecdotally is that the vast majority of tobacco farmers have their debts renegotiated. "Tobacco debts are paid with tobacco" is a common expression among tobacco farmers. According to interviews with managers of tobacco companies, fully accounting for indebtedness needs to consider the common process of renegotiating debts from one year to the next.

The goal behind the contracting strategy is, in our view, clear. The tobacco industry continually renegotiates debts to ensure that the indebted farmer remains a supplier and, typically, an exclusive one. The tobacco companies have effectively turned the "offer" of credit into a "loyalty policy" that allows players in the industry to compete for new suppliers and to keep farmers tied to them through long-term loans and contracts. This makes it easier for the companies to set national production targets within the context of a stable international market.

The repercussions of indebtedness on Brazilian farmers' lives over the long term seem to confirm observations elsewhere that tobacco farming deepens poverty (see Lecours, this volume). Bonato (2007) analyzed the Human Development Index (HDI) distribution for 2000 in tobacco-growing municipalities in the Southern region. The HDI is based on measures of nutrition status, sanitation, life expectancy, literacy rates, learning performance and per capita income, among others. The findings show that 86 percent of the tobacco-growing municipalities in Paraná (142 out of 165) have HDI levels *below* the state average. In Santa Catarina, where 251 municipalities produce tobacco, 214 of them (or 85 percent) have lower HDI levels than the state average. Finally, in Rio Grande do Sul, a state with 347 tobacco-growing municipalities, 278 or about 80 percent have a lower HDI than the state average.

The potential impacts of the tobacco industry on even poorer regions in Brazil are cause for worry. The 2006 Brazilian Agricultural Census showed that tobacco production is expanding into new areas with very few rural development assets and very low socioeconomic indicators. The development of the tobacco industry there will launch them into an agroindustrial system with very high risk of future indebtedness.

The ties created by the IPS are clearly a significant barrier when talking about livelihood diversification in tobacco-growing areas. Tobacco farmers are not truly free to make choices, especially in the absence of strong public

---

1   Souza Cruz, British American Tobacco's subsidiary in Brazil, filed a Writ of Mandamus to reserve the right to not provide 30,000 pages of information from more than 10,000 producers and obtained an injunction to withhold statements of current accounts of all producers in the state of Paraná contracted by them between 2006–2009. In September 2012 the 3rd Panel of the Federal Regional Court of the 4th Region unanimously annulled the injunction, a decision appealed by Souza Cruz.

policies for rural development and family farming. The following section examines recent efforts by the government of Brazil to create options for tobacco farmers and some of the challenges this program faces.

The National Program for Diversification in Tobacco-Growing Areas

The National Program for Diversification in Tobacco-Growing Areas (NPDTGA) emerged as a political response to competing pressures in the context of ratification of the Framework Convention on Tobacco Control (FCTC) in Brazil (Boeira and Johns 2007). Opposition to the FCTC by the tobacco industry and by political leadership in the southern states (where tobacco production is concentrated) had successfully delayed treaty ratification for almost three years, even though Brazil played a prominent role on the international stage when the treaty was being developed (Rangel 2011). Negotiations in the Senate involving six ministries eventually struck a political compromise leading to ratification. On the one hand, the coordination of the program was granted to the Technical Assistance and Rural Extension Department (DATER) in the Secretariat for Family Agriculture (SAF) in the Agrarian Development Ministry (MDA). This was to advance specific policies for family agriculture that could subsidize the diversification process. On the other hand, the program was constrained by a Statement of Interpretation of the FCTC Policy Requirements. The statement says that the program will not prohibit tobacco production or restrict access to the benefits of other national policies by those who are currently engaged in tobacco production. The statement also prevents the government from using the FCTC to engage in practices that are discriminatory to free trade. As argued below, the tensions and inherent contradictions expressed in this compromise ultimately reflect a lack of political will to break with an industry that represents a significant tax income for the government and is backed by a powerful political lobby including producers and international tobacco companies.

The Diversification Program aims to support the implementation of rural extension projects, training and research to create new opportunities for income generation in the context of rural sustainable development (Brazil 2010). The guiding principles are:

• sustainable development, to guide the development of productive and income-generating alternatives in tobacco-growing areas, committed to environmental sustainability, quality of life for families and the transition to agroecological systems;

- food security, to encourage diversified production on smallholder properties, allowing tobacco growers to have access to food on their properties of sufficient quality and quantity for consumption and sale;
- diversification, as a strategic policy action to develop local knowledge and multifunctional farms, with a focus on social, environmental and economic sustainability;
- participation, as a strategy to empower tobacco growers to autonomously set their production choices based on information and technical guidance that respect issues of gender, age and ethnicity;
- partnership, as a strategy to strengthen and broaden the process of planning and executing diversification programs with governmental and non-governmental organizations, universities, rural extension institutions, researchers and others at the national, state and municipal levels.

These principles are enabled through a variety of existing public policies aimed at strengthening specific parts of the general national food and agriculture system (Figure 7.4). The most significant for diversification are the National Program for Strengthening Family Agriculture (PRONAF) and the National Policy on Technical Assistance and Rural Extension (PNATER). Other programs aim to support market access, such as the National School Feeding Program and Family Agriculture Insurance (SEAF). Details on each of these are presented below, along with descriptions of other programs focused on specific aspects of the farming economy.

The national program for strengthening family agriculture

Public policies for agrarian development became a part of the federal government's agenda in the 1990s, expressed through the National Program for Strengthening Family Agriculture (PRONAF), housed in the Secretariat for Family Agriculture (SAF) of the Agrarian Development Ministry (MDA). The emergence of the program was a response to pressure applied since the late 1980s by rural social movements demanding agricultural credit and institutional support for small farmers excluded from existing policies. The program represented the recognition of a new social category – family farmers – who until then had been designated as small farmers, family producers, low-income producers or subsistence farmers (Schneider et al. 2004).

From the beginning the tobacco industry found a way to direct public funding from this program to tobacco farmers (Silva 2012; Bruno and Dias 2004; Abramovay and Piketty 2005). For many years the main crop funded by PRONAF was tobacco, which represented 18.42 percent of total funds released between 1996 and 2000, with corn (14.58 percent) and soybeans

Figure 7.4. Federal government policies related to the National Program for Diversification in Tobacco-Growing Areas

Source: Author, adapted from Brazil 2010.

(11.06 percent) next in line (Correa and Ortega 2002). Abramovay and Veiga (1997) emphasize that producers contracted directly by tobacco companies accounted for about half of the applications for PRONAF funding in Rio Grande do Sul and Santa Catarina. Other agroindustrial sectors were not represented as much among awarded PRONAF funds, which points to the ability of organized tobacco leaf processors and producers to draw on public funds.

In 2001, Resolution No. 2833 of the Central Bank of Brazil (Bancen) forbade granting of credit under the terms of PRONAF to the development of tobacco production in partnership with tobacco companies (Brazil 2001). Resolution No. 3559 of Bancen (2008) also changed the norms for the Rural Credit Manual (MCR) and amended PRONAF regulations along the same lines (Brazil 2008). The prohibitions were aimed at the tobacco crop itself, not at tobacco farmers. This meant that tobacco farmers were allowed access to credit for the annual costs associated with the production of other crops, even when these crops were grown alongside tobacco. The same incentives were applied to credit for capital investments. Farmers producing tobacco were allowed access to investment credit, provided that:

- The investment is not intended solely for the tobacco crop and is used in other activities that promote the diversification of farms, new crops or conversion of tobacco production units into other enterprises;
- In calculating the payment amount specified in the technical design *at least 20 percent* of the revenue generated by the production unit must originate in activities that do not relate to tobacco (Brazil 2008).

Controversy arose recently about this percentage. Resolution No. 4116 of Bancen (August 2012) increased the base percentage of revenue from 20 to 25 percent for the following year, to 35 percent in 2013–2014 and to 45 percent for the 2014–2015 season (Brazil 2012a). This dramatic change had not been previously negotiated with organizations representing family farmers and social movements or even with the tobacco-growers' associations and leaf processors. The initiative faced strong opposition from the industry, from the Agriculture Minister and from the Agrarian Development Minister (Brazil 2012b). These parties charged that the measure discriminated against smallholder tobacco farmers and would undermine the competitiveness of the tobacco supply chain (Agencia Camara de Noticias 2012). Furthermore, they argued that the resolution was made without sufficient knowledge of the importance and functioning of the sector (Gazeta do Sul 2012). They also demanded respect for the 2005 Statement of Interpretation of the FCTC committing the government to avoid practices discriminatory to free trade and to groups that engage in tobacco production (Rangel 2011; Folha de São Paulo 2012). This position was expressed directly by the Minister for Agrarian Development, Dr. Pepe Vargas, in an international seminar on Challenges for Diversification in Tobacco Growing Areas convened by the Pan-American Health Organization (PAHO) in Brasília as a preparatory event for the 5th Conference of the Parties of the FCTC (COP5). "We have to support diversification, making the transition, but we do not accept restrictive measures or reduction of the cultivated area as targets for transition," he said. Not long afterwards, Bancen issued a new resolution (No. 4136, September 2012) that reinstated the previous figure of 20 percent for the two upcoming seasons (Brazil 2012d). This was seen as a step backward for the federal policy of diversification and a decision

Photograph 7.3. Women cutting tobacco leaves to prepare dark tobacco, Arapiraca, Alagoas

Photo credit: G. E. G. Almeida.

reflecting the political influence of the entities representing the tobacco sector (O Informativo 2012).

The failure of Bancen to negotiate the change in investment credit eligibility guidelines with stakeholders prior to implementation was clearly a political blunder. Demanding a shift from 20 to 45 percent over three harvest seasons meant a rapid increase in revenue generated by activities other than tobacco. This was clearly too much to expect of farmers given the specialization inherent in the sector. As described previously, tobacco represents 96.3 percent of the total annual revenue derived by many tobacco farmers. Nevertheless, the idea underlying the eligibility criterion has merit as a means to induce new farming activities so long as it is introduced in a gradual, phased manner.

The PRONAF–Bancen controversy reflects a broader problem in how public funds to the tobacco sector are handled in Brazil and points to deep contradictions within the federal government with respect to Articles 17 and 18 of the FCTC. Between 2006 and 2012, the federal coffers provided only BRL 22.4 million to help tobacco farmers diversify their crops. This represents

6.6 percent of the BRL 336 million disbursed to tobacco agribusiness by the National Bank for Economic and Social Development (BNDES) between 2006 and 2011 – a crumb for people wanting to diversify and a banquet for the tobacco industry itself. The reasoning of the National Bank is simple. According to BNDES, there is no specific policy to encourage or discourage the tobacco sector. Loans are made to industry within a general line of credit for agriculture. Those who make a request at the right time will receive the money (O Estado de São Paulo 2012).

National policy on technical assistance and rural extension

The National Program for Diversification in Tobacco-Growing Areas (NPDTGA) is guided by the National Policy on Technical Assistance and Rural Extension (PNATER) coordinated by DATER (Technical Assistance and Rural Extension Department) in the Family Agriculture Secretariat (SAF) within the Agrarian Development Ministry (MDA). The policy aims to foster higher incomes, food security and production diversification while sustaining or creating jobs compatible with environmental and sociocultural values. Since the NPDTGA was launched in 2006, DATER has implemented more than 60 projects focused on tobacco farms, providing rural extension, capacity building and training on a wide range of topics related to agriculture and fisheries production, marketing and transformation of products and supply chains. These occurred in seven tobacco-producing states. The program reached approximately 30,000 families and 80,000 young farmers and rural youth in 600 municipalities, with a very modest investment of USD 7.1 million by 2010 (Brazil 2010).

While the principles and operational guidelines for PNATER function as guiding principles for the NPDTGA, the program also follows a sector-specific methodological framework presented by MDA in 2009 at the First Meeting of the WHO Working Group for FCTC Articles 17 and 18. According to the authors of this framework, diversification refers to a process of boosting production and employment opportunities, reducing dependency and vulnerability, enhancing quality of life, creating the underpinnings for food security and expanding the competitiveness of farmers and their involvement in inter-sector activities (Schneider et al. 2009). They argue that economically viable strategies for diversification in tobacco-growing areas should include initiatives, actions and policies centered on modifying (qualitative change) and transforming (quantitative change) the economic behaviors, cultural beliefs and ideologies imposed on tobacco farmers for decades (Schneider et al. 2009). This approach recognizes that promoting diversification beyond crop substitution also requires finding economically viable alternatives that

Table 7.3. Distinctions between crop substitution and sustainable diversification

Crop Substitution	Sustainable Diversification
Reduces dependence on the tobacco industry	Offers greater stimulus toward independence
Increases farm incomes	Diversifies farming and non-farming income
Promotes intensified resource use (land, water, labor) in production	Promotes less intensive use of productive resources
Maintains technological bases tied to the use of agrochemicals	Provides the foundation for the transition to organic and agroecological production
Contributes to reducing soil fertility and biodiversity	Fosters recovery of soil fertility and biodiversity
Promotes economies of scale and sector growth	Generates positive spin-offs for the economy
Poses risks to the health (disease) and livelihoods of producers	Fosters a greater commitment to quality of life
Fulfills the objectives of industry: continued tobacco production and supply	Fulfills public health objectives: reduced tobacco supply and consumption

Source: Adapted from Schneider et al. 2009.

contribute to reducing poverty and social vulnerability in rural areas in a sustainable manner.

The tobacco sector framework makes a clear distinction between crop substitution and sustainable diversification, as outlined in Table 7.3. This implies the promotion of rural development by enhancing the conditions that allow farmers to reduce their dependence on a single crop production system or income source. Diversification in this context means expanding scope for local economic development and creating momentum and engagement with a range of economic sectors and non-agricultural activities (Schneider et al. 2009).

The NPDTGA, and the broader PNATER within which it operates, claim to be committed to management processes and methodologies capable of supporting democratic and inclusive decision-making, including citizen engagement in planning, monitoring and evaluation of program activities (Brazil 2007). To this end, the NPDTGA organized six national seminars convening experts from MDA partner organizations, entities representing family farmers and tobacco growers, officials from government ministries, members of the National Commission for the FCTC Implementation (CONICQ) and people representing the university, research and municipal sectors.

Photograph 7.4. A young boy helps his family take dried tobacco leaves out of the oven, São Lourenço do Sul, Rio Grande do Sul

Photo credit: G. E. G. Almeida.

The people who attended the seminars, numbering almost 1000 in all between 2005 and 2010, agreed that investment in diversification is crucial for reducing producers' economic vulnerability while at the same time addressing the health problems and environmental damage linked to tobacco growing (Brazil 2010).

The program has also tried to stimulate institutional cooperation and solidarity by promoting partnerships among municipal, state and federal institutions, non-governmental organizations and small farmer organizations. Links to universities and to research groups involved in innovation and technology generation related to diversification are also encouraged, as is research on the development and assessment of the true value of local markets and insertion of farmers into global markets in ways that do not create dependence.

To support these interactions, the DATER created a Diversification Network with representation from 25 partner institutions. The goal of the network is to help manage the diversification program. So far, however, network meetings have been treated primarily as sessions to distribute information on FCTC policies to an audience that has already been steeped in the issues for many years. Critical discussions of the program and its projects, development of new activities and the formulation of demands for advancements in the implementation of FCTC Articles 17 and 18 have been limited or entirely absent from network meetings. Real evaluation and strategic planning, or even facilitation of a space for critical analysis of the program's achievements, have been lacking. According to many participants, the network exists to recharge energy, bolster hope and keep people occupied but minimally connected to decision-making. Even the network's mailing list is not set up to facilitate the direct exchange of information, articulate demands or serve as a consultative mechanism among partners. It seems to operate sporadically for official communications only, from the conveners to the members, and does not foster independent discussion. As a result, the network has not yet fulfilled its mandate of contributing directly to management of the program or insuring links with other public policies and inter-sector actions aimed at tobacco control (Brazil 2010). Nor has it become a political actor capable of demanding the structural and financial support necessary to strengthen program activities.

Coordination of the Diversification Network, and of the NPDTGA as a whole, is further disabled by the weak structural relationship of program staff to the responsible ministry. The few paid staff in the NPDTGA are temporary workers funded by the United Nations Development Programme (UNDP). Even though staff take direction from the DATER, the Director of DATER is not directly responsible for coordinating program activities, projects or research. This creates the impression that there is little or no institutional commitment to the program and its staff, and undermines its operations. Unfortunately, these kind of administrative gaps and ambiguities in the hierarchy of decision-making affect virtually all other instances where government agencies are working on matters promoted by the National Commission for the implementation of the FCTC. Governance of the process is extremely confused, making it difficult to enforce policy changes in a meaningful way.

Despite this situation, DATER has funded successful projects. When it was launched in 2005 the program received a project budget of about USD 5 million to be used during the first five years. Subsequently, the program's annual project budget fell to USD 0.5 million. With these funds DATER has financed rural extension and technical assistance actions related to rural sustainable development, food supply chains, promotion of local fairs based on economic fair trade principles, improvement and multiplication of native seeds, income generating processes and research on economic diversification. Research has been conducted on marketing opportunities, strategies for adding value to local production, agroecological food production, agroindustrial development of family farming and production, labelling, conservation and use of plants for pharmaceutical purposes.

One project with many lessons took place in the Municipality of Dom Feliciano (Rio Grande do Sul). This municipality has 2,500 farms with less than 50 hectares, 87 percent of which are family farms. The harvest produced in the municipality in 2008–2009 was mainly tobacco (86.3 percent), with some wood (9.45 percent), cattle (2.27 percent), milk (0.27 percent) and corn (0.19 percent). The economic and social challenges the municipality faces in trying to develop alternative livelihoods and promote diversification are mostly related to the high degree of specialization in tobacco farming in a region with a human development index below the state average. Around 17 percent of the population in the municipality live in poverty.

The pilot project sought to integrate federal policies and programs with regional and local approaches that enhance opportunities for rural development. Its specific goals included the creation of demonstration units for organic and free-range poultry, fish breeding and milk production. As well, the project supported the cultivation of grapes for juice and wine production. Other aspects of the project dealt with supplementary activities organized by DATER focused on income-generating processes, healthcare for tobacco farmers and communication strategies to counter misinformation spread by the tobacco industry's technicians.

To promote successful livelihood diversification, public investments were also made in Dom Feliciano through a food acquisition program (PAA) and school feeding program (PNAE), discussed below. The local mayor demonstrated a strong commitment to the goal of diversification as well. People in charge of implementing the NPDTGA also sponsored research in the municipality on the health impact of tobacco farming. Expectations of change were high, expressed as well by the mayor's strong commitment to diversification.

Unfortunately, the concentration of program attention in one municipality did not anticipate the political reality that followed. The mayor who had enthusiastically supported the pilot was not re-elected in 2012, probably due to huge financial support the tobacco industry provided to a competitor. The political battle ended in favor of tobacco production and the future of the initiatives and investments that began in Dom Feliciano are now under threat.

This experience provides further evidence of the lack of integration of diversification with Brazil's mainstream institutions. It suggests that local initiatives should be considered as components of a regional approach based on institutional cooperation and coordination at local, regional and federal levels. It is not enough to have a focused strategy based on the personal commitment of a few high-profile actors. This is particularly important given that the program is dealing with a chain of production that is well-established in more than 700 municipalities in three different states, with the economic power of the tobacco industry behind it.

Research by the Brazilian Agricultural Research Corporation (Embrapa) on the potential of different crops that grow in the country's temperate climate zones – where the country's tobacco is also grown – also underlines the importance of a broader perspective when developing programs for diversification. Table 7.4 presents partial results of the research, based on a number of different production systems in six municipalities. The research shows that potential improvements in farmers' livelihoods come from fruit production, dairy farming, honey and the processing of artisanal products. It also points to the importance of market support and direct purchasing of finished products needed to be economically viable. The project created demonstration units for production processes but also for food processing, thereby building farmers' capacity to diversify into new production areas, including fair trade markets. By contrast, the NPDTGA is only mandated to work on production, not processing or direct market support.

Program for food acquisition

The National Secretariat for Food and Nutritional Safety of the Social Development Ministry (MDS) has formulated and implemented a National Food and Nutritional Safety Policy to alleviate hunger on an emergency basis while also making structural changes to programs and projects so they can support family agriculture, regional development and food and nutritional education. In particular, the Secretariat is responsible for the coordination of the Program for Food Acquisition (PAA) run by both the MDS and MDA in partnership with state and local governments, civil society, organizations of family agriculture and social assistance networks. The PAA aims to ensure regular access to enough food for populations vulnerable to food insecurity and poor nutrition. It promotes social inclusion through the strengthening of family agriculture and helps to create inventories that allow small farmers to store their products to be sold at fair prices later.

The PAA is considered one of the main strategic actions of the Fome Zero Program (Zero Hunger), implemented through PRONAF. In summary, the program contracts the acquisition of family agriculture products at prices comparable to the regional market prices expected at harvest time. At the time of harvest, the farmer chooses whether to sell the product at this

Table 7.4. EMBRAPA research on different production systems in temperate climate zones

Name of Product	Description of research results
Citrus	Despite market fluctuations, especially in supply chains within the industry, fruit production has provided interesting economic indicators, especially for companies seeking specialized producers that constantly respond to market trends. The Southern region of Brazil is highly favored for the production of table fruits.
Manioc	This is one of the most important crops grown in Brazil, due to its many culinary uses. In several regions, it is a major source of income for farmers, especially in the production of flour and starch. It is relatively simple to produce, which has also stimulated production. Currently the root crop also figures in programs that would use this carbohydrate to produce biofuels. The project is exploring the possibility of higher incomes from manioc cultivation, building on its genetic hardiness and uniform quality.
Sweet potato	Sweet potato can be used for human consumption, including flour, as a component in animal feed and as raw material for the manufacture of alcohol. The branches are used as animal fodder. Its prowess as a cover crop is great; the plant offers excellent soil protection against the weather. The project works to offer high quality plant seedlings. The material is reproduced under laboratory conditions in order to eliminate the major diseases that affect production, enabling productivity gains in the order of 120 percent.
Figs	This fruit is well adapted to the climate of southern Brazil. Figs are mainly used for food processing, although they can be marketed fresh. Farmers grow them strategically, to supplement income. The project aims to demonstrate the technologies available for the cultivation of organic figs and to empower farmers in the setting up and managing of fig orchards.
Mini watermelon	The market for mini fruits has grown significantly in recent years in Brazil. The project is testing mini watermelon as an option for the diversification of tobacco production systems. Depending on the market differential and the simplicity of the production system, the mini watermelon may prove to be viable as an alternative to tobacco production.
Milk	Milk is a traditional product within family agriculture. Milk production on small properties can support the continuous generation of income throughout the year, and be part of other seasonal or permanent activities. The project confirms that family agriculture is primarily responsible for milk production in the southern districts. The demonstration units of milk production and grass growing have helped to transfer knowledge and technology in areas such as pasteurization. Hallmarks of the systemic changes include better milk quality, management of animal rearing and development of good habits in the act of milking.

(Continued)

Table 7.4. Continued

Name of Product	Description of research results
Honey	Honey has emerged as an important alternative to tobacco growing, with great potential for income generation on small farms. In 2007 Brazil produced about 35,000 tons of honey, the Rio Grande do Sul state being the highest producer, with 7,365 tons or 21 percent of domestic production, according to IBGE data.
Artisanal processing of food from animal and vegetable sources	The inclusion of family agriculture in the consumer market for food is recognized as a challenge. However, new perspectives are beginning to emerge. Artisanal processing of food may create the same value for production as industrialized food, considering that these products contain no chemical additives. The proposed projects aimed to reduce farmers' dependence on tobacco production, training them in the artisanal processing of food, whether of animal or vegetable origin, and thus increasing their income.

Source: Adapted from Brazil 2010.

Photograph 7.5. Burley tobacco leaves drying on a house porch, Agudos, Rio Grande do Sul

Photo credit: G. E. G. Almeida.

Table 7.5. How the food acquisition program operates

Modality	Objectives and functioning	Annual allocation of resources per family (BRL) and source	Coordination	Means of access
Direct purchase	Creates a purchase hub. In cases of low market prices or in order to meet demands from populations vulnerable to food and nutrition insecurity, this plays an important role in price regulation.	BRL 8,000 MDS and MDA	CONAB	Individuals
Storage and transformation	This supports marketing of family agriculture products, which are stored, transformed and sold when market conditions improve.	BRL 8,000 MDS and MDA	CONAB	Cooperatives and Associations
Purchase with simultaneous donation	Purchase of food produced by family agriculture. Food is donated to organizations that belong to the social assistance network.	BRL 4,500 MDS	CONAB, states and municipalities	Individuals, Cooperatives and Associations
Incentive to milk production and consumption	Support for milk consumption by families marked by food and nutritional insecurity; may help to stimulate family agriculture.	100 L milk/day BRL 1.25/L (cattle milk) BRL 1.8/L (goat milk) MDS	States from the northeast and Minas Gerais	Individuals, Cooperatives and Associations

Source: Adapted from Simoni (2009) and Brazil (2013a).

contracted price to the PAA (in coordination with CONAB, the National Supply Company) or to sell on the open market (if prices are higher). This arrangement allows advanced purchases of family agriculture production, without bidding, according to four different modalities (Brazil 2013a). Table 7.5 presents details on this program.

The PAA has introduced important innovations for family farming into the public policy arena. The guarantee of a purchase price in local markets also conserves farmers' right to sell on regional markets, thereby providing scope for the emergence of new product markets. Even so, some issues need to be addressed in the PAA design related to delivery of the program in tobacco-growing areas. Transportation logistics and sanitary regulations that restrict access to regional markets are particularly problematic for former tobacco farmers. There is also a need to strengthen the commitment of local organizations and local governments to helping farmers complete the paperwork needed to get into the program and in particular the Statement of Eligibility for PRONAF. The Statement is required to access lines of credit, public social assistance, insurance and marketing policies, among others. It is provided free of charge and issued by bodies accredited by the MDA but farmers still have difficulty getting the Statement due to differences from one ministry to another in the definition and interpretation of its provisions and requirements. A lack of interaction and coordination among ministries, and even between secretariats and departments within the same ministry, complicate the bureaucratic process enormously. Internal political disputes within each department or secretariat also create confusion and overlapping tasks. One option being considered is to unlink marketing policies like the PAA and PNAE from eligibility mechanisms used to access credit and other services, so that a wider range of small farmers can access markets. These adjustments are critical to overcoming a key constraint on diversification among tobacco farmers: access to a range of markets.

National school feeding program

The National School Feeding Program (PNAE) ensures the feeding of pupils receiving basic education in public schools and from charities. The goal is to meet the nutritional needs of students to contribute to their learning and performance, as well as to promote healthy eating habits. Despite more than 50 years of existence, the legal framework for the PNAE was only sanctioned in 2009, thanks primarily to the work of the National Council for Food and Nutrition Security (CONSEA) and broad-based mobilization of civil society organizations. Dispute over sanction of the program raged in the Senate due to the strength of the private sector within the food industry and the rural caucus that tried to monopolize the institutional market for school feeding. The new law recognizes food as a human right and sets out the obligation that at least 30 percent of the funds it provides will buy food from family farms through public calls for purchase, without bidding.

According to the National Foundation for Educational Development (FNDE), the agency responsible for the program, the federal government provides states and municipalities BRL 0.30 to BRL 1.00 per pupil for each school day. These figures are based on the students' level of education and calculated based on school census data from the previous year. Monitoring and oversight of the program occurs through councils, the court of auditors and prosecutors and other institutions. The 2012 budget for PNEA was BRL 3.3 billion for the benefit of nearly 45 million students. From this total, approximately BRL 990 million was to be used to buy directly from family farmers. The 2013 budget contains estimates of about BRL 3.5 billion in total, with about BRL 1 billion hallmarked for payments to family farmers (Brazil 2013b).

The acquisition of food is provided, whenever possible, in the counties where the schools are located. When supplies cannot be obtained locally, schools can complete the demand by turning to farmers from the same region, rural area, state and country, in that order. Until July 2012, an individual farmer could sell up to BRL 9,000 per year to the program, which FNDE Resolution 25 increased recently to BRL 20,000. This increase in the upper limit may stimulate more farmers to participate in the program.

In some tobacco-producing regions, PNAE has not reached the minimum target of allocating 30 percent of its purchasing from family farmers, mostly because food production does not meet the variety and quantity criteria required by the program. Given the high degree of specialization among tobacco farmers, providing the right products in the right amount is difficult. Nevertheless, the policy has the potential to gradually stimulate food production in tobacco-growing areas by guaranteeing an institutional market for fresh and processed local foods for use in schools.

Other policies for family agriculture

While the policies and programs described above constitute the main thrust of the National Program for Diversification in Tobacco-Growing Areas (NPDTGA), other policy and program instruments related to strengthening family agriculture also help in specific ways. The Family Agriculture Insurance (SEAF) is an action directed exclusively at family farmers who use credit from PRONAF for operational costs. SEAF was established under the Program for the Guarantee of Agricultural Activities (PROAGRO), which responds to a longstanding farmer demand: crop insurance. PROAGRO partly covers repayments on rural credit used for operational costs when crop harvests and livestock are affected by pests and diseases. The insurance guarantees 65 percent of the net revenue estimated when the operational cost credit is calculated.

The Price Guarantee Program for Family Agriculture Products (PGPAF) ensures a discount on loan payments in situations where market prices fall below prices anticipated when production costs are calculated and loans provided.

Meanwhile, the Harvest Guarantee Program is an insurance program for producers living in the semi-arid regions of Brazil. Farmers who join this program are compensated when there are proven losses of at least 50 percent for crops such as cotton, rice, beans, cassava and corn. Tobacco farmers may use these programs to help them diversify.

Sustainable PRONAF is another credit instrument that is also relevant to the challenges of diversification among family tobacco farmers. It was created in 2008 to provide farmers with access to technical assistance for farm planning with a focus on compliance with environmental standards and the use of sustainable production practices that also increase productivity and farm income. It involves a participatory process to review past farming practices, income and cash flow. This helps to develop a long-term vision for farms and farm businesses (Brazil 2010). However, infrequent inspection and weak controls due to the lack of human resources in the responsible ministry have led to highly visible cases of fraud. Some tobacco farmers accessing the program for planning crop diversification were found to be using the credit to improve their tobacco-growing operations.

Conclusion

Brazil's approach to promoting economic and productive alternatives to tobacco through the National Program for Diversification in Tobacco-Growing Areas (NPDTGA) has two main features. First, the program has firmly rooted the transition challenge in the context of broad policies and programs that address problems facing family farms. It has not introduced special policies and programs of its own but rather provided a conceptual framework for coordinating and facilitating access in tobacco regions to *existing* policies for family farming. This creates a political space for dialogue and action on a wide range of issues facing family farms, including access to credit and technical assistance, research on smaller scale production systems, farm financing, the organization of markets suited to family farms, fiscal and tax incentives, etc. The approach elevates the debate about alternatives to tobacco from a search for crop substitutes to questions of structural constraints facing family farms and the politics of competitiveness in the agricultural sector.

Second, the program focuses on livelihood diversification as a worthy end in and of itself and a general political goal. The harmful effects of dependence on a single production chain are particularly evident in the tobacco sector, but not limited to this sector. By promoting a discussion on the socioeconomic vulnerabilities of tobacco farmers the program has helped to prompt a debate on farm specialization in Brazil and to promote research and training on diversification as a key element aimed at overall improvement of the sector. Prior to the program, the Agrarian

Development Ministry (MDA) had not considered the question of diversification in the context of tobacco farming or any other monoculture. PRONAF, in many cases, had actually served to deepen specialization of family agriculture. The recent resolution of Bancen, despite its political failure, is a bold example of what a coherent approach to agricultural diversification could look like. It shows that stimulating income diversification through a credit access policy has the potential to stimulate new products and farming activities. Where it failed was the lack of consultation and cooperation with the farming base.

The emergence of these two features in the NPDTGA (transition in the context of development of the family farm and diversification as a general benefit in agriculture) represents a positive political gain. Actual impact on the ground, however, has been quite limited. The public system of DATER does not have enough human, technical or financial resources to respond to general demand from farmers, let alone the special needs of tobacco farmers. Weaknesses such as a lack of coordination between different policies and the low level of investment in infrastructure, financial resources and human resources dedicated to the program continue to weaken its effectiveness. Furthermore, many key stakeholders in favor of family agriculture (governments, rural labor unions, social movements, churches and non-governmental organizations) are reluctant to move against tobacco growing. While they would like to have better policies for rural development in their regions, they have accepted the industry argument that a transition out of tobacco growing will reduce current socioeconomic standards. Municipalities in key tobacco-growing states such as Rio Grande do Sul have been particularly vocal in their opposition to diversification strategies that do not allow for tobacco production. They have long-term partnerships with tobacco industries and have expressed little interest in the regional approach to local economic development needed for a successful process of farm diversification. There are many interests within the federal government as well that do not see family farming as a priority component of rural development strategies. Without the support of these policy actors, most farmers cannot independently invest time and other limited resources in alternative crops or develop non-farming activities. Many prefer to continue with their practice of tobacco growing. While cooperatives have the potential to overcome this inertia by organizing farmers and scaling up collective action, the high tax burden applied to cooperatives in Brazil has so far undermined their use as part of the diversification program.

While providing scope for some change, the official discourse on diversification has so far avoided and even banned debate regarding the longer-term future of tobacco growing in Brazil. The notion of *supply reduction* currently has no place in the government's vocabulary. It is not even an acceptable long-term target for rural development policies, even within a new territorial approach (Brazil 2012c). Considering the existing strong demand for tobacco leaf in the international market, it is feasible to imagine that the Integrated Production System of Brazilian tobacco will continue to grow and spread to new areas, reaching a larger number of farmers than today, each of them cultivating small tobacco plots. If farmers currently involved in tobacco farming in Southern Brazil decide to take advantage of government support mechanisms and diversify their production by diminishing the cultivated area under tobacco on their farms, tobacco industries and leaf processors will simply focus on new growers in the Northeast, where there are economically and social vulnerable family farmers. Even in Southern Brazil there are many potential growing areas that have not yet been tapped by the tobacco companies. To be truly coherent with the goals of the diversification program, a fair and orderly exit from all tobacco production must eventually be considered.

In our view, a commitment to diversifying family farm livelihoods should include the resolve to limit the cultivation of tobacco to those areas where it is already grown. Diversifying without preventing the shift of national production to other areas and restricting the entry of new farmers into the IPS will simply shuffle the social, economic and environmental threats around within the country. It will also strengthen the tobacco industries' political position by appearing to cooperate with the common goal of diversification. This scenario could be avoided, however, if the reduction of tobacco growing in the country as a whole were seen as a natural consequence of territorial development (Rocha and Bursztyn 2007). Territorial development considers opportunities and comparative advantages that cut across administrative boundaries or localities and even regions, providing not only a new scale for processes of development but also a new method for favoring those processes (Miranda and Costa 2007). Currently, recognition of this approach in relation to tobacco farming is rejected by a government that insists that the reduction of tobacco growing is not an acceptable target.

Brazil has the skills, methods, legislative frameworks and other resources needed to implement a broad territorial development strategy that offers economic alternatives to tobacco production and real scope for farmers to make choices. What is needed is to follow the diversification program through to its logical conclusion by promoting agrarian development that is people centered and territorially based.

References

Abramovay, R. and J. E. Veiga. 1997. "Novas Instituições para o Desenvolvimento Rural: o caso do Programa Nacional de Fortalecimento da Agricultura Familiar (PRONAF)." Discussion paper published by IPEA (Texto para Discussão n. 641). Online: http://www.ipea.gov.br/pub/td/1999/td_0641.pdf (accessed 15 October 2012).

Abramovay, R. and M. G. Piketty. 2005. "Política de crédito do Programa Nacional de Fortalecimento da Agricultura Familiar (Pronaf): resultados e limites da experiência Brazileira nos anos 90." *Cadernos de Ciência & Tecnologia* (22)1: 53–66.

AFUBRA (Associação Dos Fumicultores Do Brazil). 2012. "Fumicultura no Brazil – Distribuição Fundiária." Online: http://www.afubra.com.br/index.php/conteudo/show/id/80 (accessed 16 October 2012).

_____. 2013. "Evolução da Fumicultura no Brasil." Online: http://www.afubra.com.br/index.php/conteudo/show/id/83 (accessed 19 April 2013).

Agência Câmara De Notícias. 2012. "Produtores de fumo reclamam de restrições no acesso ao crédito do Pronaf." Online: http://www2.camara.leg.br/agencia/noticias/agropecuaria/422699-produtores-de-fumo-reclamam-de-restricoes-no-acesso-ao-credito-do-pronaf.html (accessed 15 October 2012).

Agostinetto, D., L. E. A. Puchalski, R. de Azevedo, G. Storch, A. J. A. Bezerra and A. D. Grützmacher. 2000. "Caracterização da fumicultura no Município de Pelotas." *Rev. Bras. de AGROCIÊNCIA* 6(2): 171–75.

AGROSTAT. 2012. "Estatísticas de Comercio Exterior do Agronegócio Brazileiro." Online: http://extranet.agricultura.gov.br/primeira_pagina/extranet/AGROSTAT.htm (accessed 15 October 2012).

Almeida, G. E. G. 2005. *Fumo: Servidão Moderna e Violações de Direitos Humanos.* Curitiba: Terra de Direitos.

Beling, R. 2003. *Anuário Brasileiro de fumo 2003.* Santa Cruz do Sul: Gazeta.

Boeira, S. and P. Johns. 2007. "Indústria de Tabaco vs. Organização Mundial de Saúde: um confronto histórico entre redes sociais de *stakeholders*." *Revista Internacional Interdisciplinar Interthesis,* 4(1): [no page numbers]. Online: https://periodicos.ufsc.br/index.php/interthesis/article/viewFile/895/10851(accessed 18 December 2013).

Bonato, A. 2007. *Perspectivas e desafios para a diversificação produtiva nas áreas de cultivo de fumo – a realidade da produção de fumo na região Sul do Brazil.* Curitiba: DESER.

Bonato, A., C. F. Zotti and T. de Angelis. 2010. *Tabaco: da produção ao consumo, uma cadeia da dependência.* Curitiba: DESER/ACT/HealthBridge.

Brazil. 2001. "Resolução n. 2.833, de 25 de abril de 2001. Brasília: Banco Central do Brazil." Online: http://www.bcb.gov.br/pre/normativos/res/2001/pdf/res_2833_v2_L.pdf (accessed 15 October 2012).

_____. 2007. "Política Nacional de Assistência Técnica e Extensão Rural." Report published by MDA/SAF/DATER, Brazilia, 22 pp.

_____. 2008. "Resolução n. 3.559, 28 de março de 2008. Brasília: Banco Central do Brazil." Online: http://www.bcb.gov.br/pre/normativos/res/2008/pdf/res_3559_v2_P.pdf (accessed 15 October 2012).

_____. 2010. "Actions of the Ministry of Agrarian Development for the Diversification of Production and Income in Areas of Tobacco Cultivation in Brazil." Report published by Ministério do Desenvolvimento Agrário.

_____. 2012a. "Resolução n. 4116, de 02 de agosto de 2012. Brasília: Banco Central do Brazil." Online: http://www.bcb.gov.br/pre/normativos/res/2012/pdf/res_4116_v1_O.pdf (accessed 15 October 2012).

_____. 2012b. "MDA busca manter em 20% exigência da receita de outras atividades para produtores de fumo." [Press release] Ministério do Desenvolvimento Agrário, 8 August. Online: http://portal.mda.gov.br/portal/noticias/item?item_id=10327103 (accessed 18 December 2013).

_____. 2012c. "Pepe destaca importância da diversificação da produção do tabaco."[Press release] Ministério do Desenvolvimento Agrário, 10 October. Online: http://portal.mda. gov.br/portal/saf/noticias/item?item_id=10579658 (accessed 18 December 2013).

_____. 2012d. "Resolução 4.136, de 27 de setembro de 2012. Brasília: Banco Central do Brazil." Online: http://www.bcb.gov.br/pre/normativos/res/2012/pdf/res_4136_ v1_O.pdf (accessed 15 October 2012).

_____. 2013a. "Aquisição de Alimentos e Promoção Social. Brasília, Ministério do Desenvolvimento Social." Online: http://www.mds.gov.br/segurancaalimentar/ decom (accessed 18 April 2013).

_____. 2013b. "Apresentação – Programa Nacional de Alimentação Escolar." Announcement by Fundo Nacional de Desenvolvimento da Educação. Online: http:// www.fnde.gov.br/programas/alimentacao-escolar/alimentacao-escolar-apresentacao (accessed 18 April 2013).

Bruno, R. A. L. and M. M. Dias. 2004. "As políticas públicas de crédito para os assentamentos rurais no Brazil." Report published by Relatório de Consultoria, Rio de Janeiro, 74 pp.

Buainain, A. M. and H. M. de Souza Filho (ed.). 2009. *Organização e funcionamento do mercado de tabaco no Sul do Brazil*. Campinas: Unicamp.

Buckles et al., this volume.

Carvalho, H. M. 2005. *O campesinato no século XXI – possibilidades e condicionantes do desenvolvimento do campesinato no Brazil*. Petrópolis: Vozes.

Chaaban, this volume.

Chonchol, J. 1986. *Sistemas agrarios en América Latina: de la etapa prehispánica a la modernización conservadora*. Santiago del Chile: Fondo de Cultura Economica.

Corrêa, V. P. and A. C. Ortega. 2002. "PRONAF – Programa Nacional de Fortalecimento da Agricultura Familiar, qual o seu real objetivo e público-alvo?" In Sociedade Brasileira de Economia Rural, Proceedings of XL Congresso Brasileiro de Economia e Sociologia Rural,Vol. 40, Passo Fundo/RS.

Eriksen, M., J. Mackay and H. Ross. 2012. *The Tobacco Atlas*, Fourth Edition. New York: World Lung Foundation.

FAO (Food and Agriculture Organization). 2003. *Issues in the Global Tobacco Economy: Selected Case Studies*. Rome: Commodities and Trade Division, Food and Agriculture Organization.

Folha De São Paulo. 2012. "Aumenta restrição a crédito para produtores de fumo." Mercado. 8 July. Online: http://www1.folha.uol.com.br/fsp/mercado/53363-mercado-aberto. shtml (accessed 10 December 2012).

Furtado, C. 2000. *Formação Econômica do Brazil*. São Paulo: Companhia Editora Nacional and Publifolha.

Gazeta Do Sul. 2012. "Fumo: governo precisa cumprir acordo!" *Gazeta Do Sul*, 10 August. Online:http://www.gaz.com.br/gazetadosul/noticia/361901-fumo_governo_ precisa_cumprir_acordo/edicao:2012-08-10.html (accessed 10 December 2012).

IBGE (Instituto Brazileiro de Geografia e Estatística). 2010a. *Sistema IBGE de Recuperação Automática – SIDRA*. Brazil: IBGE. Online: http://www.sidra.ibge.gov.br (accessed 14 August 2012).

_____. 2010b. "Produção Agrícola Municipal Culturas Temporárias e Permanentes. Vol. 37." Report published by IBGE, Rio de Janeiro, 91 pp.

Iglesias, R. and A. Biz. 2011. "Análise da fumicultura e relação com política de controle do tabagismo no Brazil." Fact sheet published by ACTbr, Rio de Janeiro, 5 pp.

Lecours, this volume.

Miranda, C. and C. Costa. 2007. *Ações de Combate à Pobreza Rural: metodologia para avaliação de impactos*. Série Desenvolvimento Rural Sustentável. Brasília: IICA.

O Estado De São Paulo. 2012. "Indústria do fumo toma R$ 336 mi do BNDES em 5 anos." *Estado*, 9 September. Online: http://economia.estadao.com.br/noticias/economia+geral,industria-do-fumo-toma-r-336-mi-do-bndes-em-5-anos,125965,0.htm (accessed 20 October 2012).

O Informativo. 2012. "Banco Central cede à pressão e reverte medidas do Pronaf." *O Informativo*, 3 October. Online: http://www.informativo.com.br/site/noticia/visualizar/id/27933/?Banco_Central_cede_a_pressao_e_reverte_medidas_do_Pronaf.html (accessed 20 October 2012).

Prado Jr., C. 2000. *Formação do Brazil Contemporâneo: Colônia*. São Paulo: Braziliense; Publifolha, 149 pp.

Rangel, E. C. 2011. *Enfrentamento do controle do tabagismo no Brazil: O papel das audiências públicas no Senado Federal na ratificação da convenção quadro para controle do tabaco (2004–2005)*. Rio de Janeiro: Fiocruz.

Rocha, J. D. and Bursztyn, M. 2007. "Políticas Públicas Territoriais e Sustentabilidade no Semi-árido Brazileiro: a busca do desenvolvimento via Arranjos Produtivos Locais." In proceedings from Anais do VII Encontro da Sociedade Brazileira de Economia Ecológica, Fortaleza, 28–30 November.

Schneider, S., L. Mattei and A. A. Cazella. 2004. "Histórico, caracterização e dinâmica recente do Pronaf – Programa Nacional de Fortalecimento da Agricultura Familiar." In *Políticas Públicas e Participação Social no Brazil Rural* edited by S. Schneider, M. K. Silva, and P. E. M. Marques, 21–50. Porto Alegre: UFRGS Editora. Online: http://www.ufrgs.br/pgdr/livros/serie_estudos_rurais/18.pdf (accessed 19 December 2013).

Schneider, S., M. Perondi and A. Gregolin. 2009. *References for the Development of Economically Viable Alternatives to Tobacco Production and the Diversification of the Livelihoods of Farmers*. Brasília: MDA.

Silva, S. P. 2012. "Políticas públicas, agricultura familiar e desenvolvimento territorial: uma análise dos impactos socioeconômicos do Pronaf no território Médio Jequitinhonha - MG. Texto para Discussão n. 1693." Report published by IPEA, Brazilia, 40 pp. Online: http://www.ipea.gov.br/portal/index.php?option=com_content&view=article&id=15139 (accessed 19 December 2013).

Silveira, R. L. L. and M. Dornelles. 2010. "Mercado Mundial de tabaco, concentração de capital e organização espacial. Notas introdutórias para uma geografia do tabaco." *Scripta Nova. Revista Electrónica de Geografía y Ciencias Sociales* 14(338): 741–98. Online: http://www.ub.edu/geocrit/sn/sn-338.htm (accessed 12 December 2012).

Simoni, J. 2009. "A multidimensionalidade da valorização de produtos locais: implicações para políticas públicas, mercado, território e sustentabilidade na Amazônia." Tese de doutorado [PhD diss.], Universidade de Brasília.

Sinditabaco (Sindicato Interestadual da Indústria do Tabaco). 2012. Brazil: SINDITABACO. Online: www.sindifumo.com.br (accessed 30 June 2012).

Vargas, M. A., N. D. S. Fihlo and R. M. Aleivi. 1998. "Análise da dinâmica inovativa em arranjos produtivos locais no RS: complexo agro-industrial fumageiro." Paper published by Universidade de Santa Cruz do Sul (UNISC), Centro de Estudos e Pesquisas Econômicas (CEPE), 49 pp.

Vargas, M. A. and R. R. Campos. 2005. "Crop Substitution and Diversification Strategies: Empirical Evidence from Selected Brazilian Municipalities." Discussion Paper, Economics of Tobacco Control, No. 28 published by The World Bank, Washington, 50 pp.

Vargas, M. A. and B. Oliveira. 2012. "Estratégias de Diversificação em Áreas de Cultivo de Tabaco no Vale do Rio Pardo: uma análise comparativa." *Revista de Economia e Sociologia Rural* 50(1): 175–92.

WHO (World Health Organization). 2008. "Study Group on Economically Sustainable Alternatives to Tobacco Growing (in relation to Articles 17 and 18 of the Convention)." Paper presented at Conference of the Parties of the Framework Convention on Tobacco Control (Third session), Durban, South Africa, 17–22 November. Online: http://apps. who.int/gb/fctc/PDF/cop3/FCTC_COP3_11-en.pdf (accessed 9 April 2013).

World Bank. 1999. "Curbing the Epidemic: Governments and the Economics of Tobacco Control." *Tobacco Control* 8: 196–201.

Conclusion

REFRAMING THE DEBATE ON TOBACCO CONTROL AND TOBACCO FARMING

Daniel Buckles, Natacha Lecours and Wardie Leppan

Introduction

From a public health perspective the case for tobacco control is compelling. Policies to reduce tobacco use have been successfully implemented in many contexts, leading to improved health outcomes for all segments of society (Centers for Disease Control and Prevention 2007; Glantz and Gonzalez 2011; Drope 2011). The fiscal health of national economies is also positively affected by reducing the prevalence of domestic tobacco use. While the size of the net benefit to the economy depends on a variety of domestic factors (Warner 2000), there is no doubt that it is positive at present and only going to grow more positive over time as populations and the economic costs of tobacco-related disease and death increase. Policies aimed at reducing the prevalence of tobacco use (taxation, smoking bans, education, etc.) also bring net economic benefits to individual users of tobacco products and the broader economy as expenditures shift to productive uses (Roy et al. 2012; Jha and Chaloupka 1999; Warner et al. 1996; Townsend et al. 1994). These gains in knowledge and practice are encouraging, and have led some tobacco-control experts to begin to treat the idea of ending the tobacco epidemic as an attainable goal.[1]

Despite the unarguable merits of tobacco control, implementation of the Framework Convention on Tobacco Control (FCTC) is only just beginning

---

[1] A special issue of Tobacco Control (2013, Volume 22) describes and assesses a range of strategies for ending the tobacco epidemic.

in many countries. While delays are due to several factors, including the complexity of the policy making environment and the challenges of enforcement, the influence of the tobacco industry on governments and policy makers has been a significant limiting factor (Malone et al. 2012; Callard and Collishaw 2013; Drope 2011; Otañez et al. 2009). Strategies used for decades by the tobacco industry to dilute, delay and defeat tobacco control in high-income countries (HIC) are now being redeployed successfully in low- and middle-income countries (LMIC) where most of the world's tobacco is now grown and where growth in consumption is greatest (Lee et al. 2012; Jones et al. 2008). Much like climate change deniers and defenders of the asbestos industry, tobacco company supporters seed uncertainty and debate where there should be none.

Prominent among current tobacco industry tactics are claims of solidarity with the world's tobacco farmers even though the tobacco industry and farmers have more competing than common interests. Tobacco is a vertically integrated industry that stimulates over-production of tobacco leaf at the farm level and uses its monopoly power to drive down farm-gate prices (see below and case studies, this volume). Contract farming, farmer indebtedness and under-grading of tobacco leaf are among the more powerful tools used to achieve this. Industry claims that the fate of tobacco farmers is tied to the fate of the tobacco corporations hides this divergence of interests and undermines the development of economically sustainable farming alternatives. The tobacco industry also plants and perpetuates myths, half-truths and fabrications concerning the immediate economic consequences of tobacco control for tobacco farmers and attempts to cast farmers in the role of victims of tobacco-control policy. These tactics, used to great effect in the USA (Benson 2011), are even more troubling in the context of LMICs where tobacco farming also contributes to environmental degradation, the use of child labor and food insecurity (see Lecours, this volume).

The chapters of this book take on various dimensions of the myths perpetuated by the tobacco industry. They do so by examining three specific questions – the determinants of demand and prices for tobacco leaf, the harsh realities of tobacco farming in LMIC contexts and practical experiences with the transition to economically sustainable alternatives to tobacco production. This chapter synthesizes and broadens lessons from the research and outlines a clear and positive position that governments, public health specialists and civil society organizations can take with respect to tobacco control and tobacco farming in LMICs. In doing so, it seeks to reframe the debate on tobacco control and tobacco farming away from its current reactive and defensive stance. Clearly, there is no reason to be concerned about the impact of domestic tobacco-control policies on farmer livelihoods in a context where global demand for tobacco leaf continues to remain strong. The reality is that governments can act with a resolve that does not pit tobacco control against the current generation of tobacco farmers. This book shows why this is the case and begins to outline agricultural policy reforms and programs that can enable farmers to begin the transition to economically sustainable alternatives to tobacco.

Drivers of Demand and Prices for Tobacco Leaf

Global forecasts for cigarette consumption, the tobacco product that accounts for most uses of tobacco leaf, indicate steady growth well beyond 2020 (Eriksen et al. 2012; Chaaban, this volume). As the vast majority of tobacco farmers around the world produce a product that is traded on this global market, they are largely unaffected by demand reduction measures in their own countries. The drivers of demand and prices for tobacco leaf, a central concern of tobacco farmers, lie elsewhere and are grounded in the business model of the tobacco industry. The drivers include:

- significant scope for new customers;
- the search for the lowest labor costs;
- the mining of forests and soil resources;
- vertical integration; and
- declining public investment in agriculture.

Research on the evolution of the tobacco industry in Lebanon, Malawi, Bangladesh, Kenya and Brazil presented in this book illustrates how these factors and actors work together in each context to create dependency and exploitation of tobacco farmers and tobacco-producing communities along the way.

Scope for new customers

Demand for tobacco leaf is driven first and foremost by keeping current customers for tobacco products and the successful recruitment of new customers through aggressive marketing. Population increase globally provides scope for significant growth in the consumption of manufactured tobacco products and consequently demand for tobacco leaf (Chaaban, this volume). Most of the growth in consumption is in Asia, where rising incomes and limited awareness of the harmful effects of smoking have created fertile ground for the recruitment of new customers (Eriksen et al. 2012). The Middle East is also experiencing significant rates of increase in the use of waterpipe tobacco smoking among youth (Nakkash et al. 2011). Urbanization, rising income and marketing of a western image are also bringing the tobacco epidemic to new heights in parts of Africa, adding to the burden of disease and death. Globally, the absolute number of users of tobacco products continues to grow through sales to new customers, and in particular to youth (Hipple et al. 2011) and women. Even in countries where tobacco-control measures have been strongest, the initial decline in the use of tobacco products has leveled off at a steady state of current customers. The global demise of the cigarette and other forms of tobacco use, long predicted and long desired, is very slow in coming and by no means certain.

Lowest labor costs

Tobacco is an introduced crop everywhere in the world except North America.[2] While disseminated widely during the nineteenth and twentieth centuries, until recently much of the worlds' commercial tobacco was grown in the United States, Canada, Australia and several European countries. Today, most commercial tobacco is grown in LMICs. This shift began in the 1960s and reached 90 percent of global production by 2006 (Geist et al. 2009).

The shift of tobacco leaf production from the North to South reflects the globalization of input supply underlying the evolving business model of the big industry players – the five or so multinational corporations that manufacture the vast majority of cigarettes consumed worldwide (Eriksen et al. 2012). The business principle is the search for the lowest cost when meeting business needs, which in tobacco production means the labor required to grow, harvest and dry the tobacco leaf purchased by the wholesale market. This is different from industries where labor is a small and diminishing fraction of total costs. All of the case studies presented in this book, and many others published elsewhere, point to the labor intensive nature of tobacco farming and the overriding interest of multinational corporations to seek out cheaper labor wherever it can be found. This includes child labor and countries with governments that do not enforce child labor laws (Amigó 2010; Otañez et al. 2006). This geographic reorganization of tobacco production finds its origins in structural changes to the global economy prompting the formation of the "world market for labor" (Oluwafemi 2012) and a world market for raw materials and industrial inputs.

Mining of forests and soils

The business model of the tobacco industry also revolves around the mining of forest resources critical to the supply of flue-cured tobacco (Lecours, this volume).

---

2   Tobacco was and still is used by the aboriginal peoples of North America that
    domesticated tobacco as a sacred plant.

The situation in Bangladesh described in this book (Akhter et al.) illustrates the strategy, which involves shifting production from one region to another within the country as fuel sources (and soils) are exhausted. The most recent shift involves the rapid expansion of the flue-cured tobacco industry into Bandarban in southern Bangladesh where the forests of the Chittagong Hill Tracts are still abundant and the soils along the Matamuhuri River are replenished each year by shoreline flooding. Environmental regulation and enforcement in Bangladesh has not kept up with the more agile strategy of the tobacco industry. Nor has investment in meaningful reforestation, leaving behind denuded hillsides and degraded soils.

Vertical integration

Vertical integration refers to control of the supply and marketing chain through a common owner. In many tobacco-growing nations multinational tobacco manufacturers have established subsidiaries that operate as wholesale buyers of tobacco leaf and establish direct relationships with farmers. For example, British America Tobacco, which describes itself on its website as one of the world's most international businesses, operates 44 cigarette factories in 39 countries. In many of these countries it is also a prominent wholesale buyer of tobacco leaf, either directly or through associate companies. The company, according to its website, has more than 140,000 contracted farmers and many more through informal networks of local, small-scale tobacco traders. This "backward" vertical integration (from the final product backwards to the raw materials) brings the supply of the main input used in the production of cigarettes (tobacco leaf) under corporate control. It enables cigarette manufacturers to reach further down the value chain into the practice of farming, thereby ensuring the predictable supply of raw materials for cigarette manufacture, in quantities, qualities and varieties when and where they are needed. Control is established without having to assume the costs of land ownership or risk at the farm level.

Vertical integration also provides scope for shifting profit centers from one part of the value chain to another. As discussed by Hamade (this volume), tobacco companies are willing to pay more than the world price for tobacco grown in Lebanon because their presence in the country allows them to sell much larger quantities of manufactured tobacco products into the Lebanese and illegal regional markets. This logic is similar to that found in the agrochemical industry where seed production by one part of the company can operate at a loss so long as farmers are required through contracts to purchase pesticides from another part of the company that has much higher corporate profit margins (ETC Group 2013).

Intermediate buyers, government agencies and farmers associations, at one time players in national tobacco leaf markets, are increasingly subordinate to a handful of vertically integrated multinational tobacco corporations (Lee et al. 2012). Farmers and local tobacco traders also play a subordinate role to a few buyers at the top of the chain. Where tobacco auction houses are still in place, as in Malawi, vertically integrated tobacco corporations actively lobby governments to restructure the industry to allow them direct contact with tobacco farmers (Otañez and Graen, this volume). As noted by Otañez and Graen, the 2012–13 growing season in Malawi marked a watershed year for the country as contract farming that year accounted for 80 percent of total trade by volume when previously auction houses had dominated. When combined with market monopolies dominated by very few buyers, contract farming inevitably weakens the bargaining position of farmers and increases their vulnerability to price fixing and downgrading of product. This has quickly become the rule in the tobacco industry worldwide.

Declining public investment in smallholder agriculture

The extent to which tobacco farming appeals to farmers depends on what other options are available to them. As can be seen from the case studies presented in this book, the evolution of tobacco farming in low- and middle-income countries took advantage of global political and economic forces that weakened smallholder agriculture and allowed tobacco farming to advance in an agricultural policy vacuum. Structural adjustment imposed on LMICs as a condition for receiving loans resulted in drastic cuts to government investment in food-focused agriculture by smallholders and a shift to export crops favoring economies of scale and the lowest price on the global market. A recent assessment of the impact of the World Bank's reforms in Sub-Saharan Africa (Stein 2010) found no improvement, however, in the agricultural sector after years of structural adjustment. The study attributed the failure of the policies in large part to a dramatic decline in government expenditure on agriculture in the post-adjustment period. According to the study, "Government expenditures went from 20 percent of GDP in the pre-adjustment period of the 1970s to 14 percent in the 1990s and down to 13 percent by 2006. [...] African countries investment in irrigation systems and extension and research collapsed in the 1980s and 1990s (Stein 2010, 5)."

The aim of the structural adjustments policies of the 1980s and 1990s was to reduce national deficits and redirect government attention to international trade (Tobin and Knausenberger 1998). The programs have been widely criticized, however, for focusing development on a few commodities where international competitiveness is determined by the lowest price for raw materials (George 1988). They also undermined long-term development of the agricultural sector. The sharp decline in fertilizer use in Sub-Saharan Africa and collapse of government marketing systems following structural adjustment reforms resulted in dramatic reductions in per capita food production. For example, by 2006 "Tanzania food production per capita was 30 percent below the pre-adjustment level of 1986 (Stein 2010, 7)."

In Malawi, reforms to government agricultural policy and spending sought to liberalize the overall production and marketing environment for cash crops, with particular attention to the production of burley tobacco (Otañez and Graen, this volume). Aimed at promoting export crops, Bangladesh's crop "diversification" programs of the 1980s and 1990s situated tobacco among the cash crops supported by international aid and government programs, a privileged position it continues to occupy today despite its new pariah status (Akhter et al., this volume). In Kenya, tobacco was embedded in the agricultural landscape through colonial powers. Subsequent government investment in tobacco exports brought further commercial advantage to the industry while as the same time deflecting policy attention away from food production and smallholder agricultural development (Kibwage et al., this volume).

The ideas promoted by international financial institutions that pushed for specialized export crops and the reduction of government expenditure are still very present in the way that LMICs govern their economies, including a tendency to under-invest in smallholder and domestically focused and broad-based agriculture. Under these conditions, markets for agricultural products and market infrastructure (petty traders, storage and aggregation facilities, transportation services, product transformation) have withered, making it difficult to sell traditional food crops, add value to basic agricultural goods or meet demand for new products that could challenge tobacco. Without access to markets and market infrastructure, farmers have been left with few meaningful choices. The tobacco industry's promise of cash, extension support and ready buyers has filled this vacuum, despite the poor terms of trade and economic dependency that comes with it. Any transition strategy aimed at providing farmers with economically sustainable alternatives to tobacco needs to be placed in this context of a generalized decline in smallholder agriculture and the absence of government investment in broad-based agricultural development.

The Elements of a Tobacco-Farming Transition

While global demand for tobacco leaf is expected to remain relatively stable for years to come, there are many good reasons why governments should pursue agricultural policy reforms and programs that create the conditions for economically sustainable alternative crops and rural livelihoods. As documented in this volume by Lecours, and amply illustrated by the case studies in this book, tobacco farming in LMICs as it is practiced today does not lift farmers out of poverty. The profitability of tobacco farming quickly evaporates into losses once even the most basic costs such as labor and inputs are factored into the calculation of net gains. The system is sustained in many contexts by the accumulation of debt, the exploitation of child labor and the selling of false hopes for seasons to come, not income and savings normally expected from a viable farming operation. Meanwhile, many other costs such as exposure to health hazards and environmental degradation are borne by tobacco-farming families, field workers and neighboring communities. The worldwide tobacco pandemic is rightly seen not only as an important health issue but also a cause and effect of a general crisis in smallholder agriculture. Creating the conditions for a transition must consequently consider how tobacco-farming fits in the broader system of agricultural land use and agricultural policies and programs. It also implies engaging tobacco farmers and other stakeholders in shaping the pace and direction of a tobacco-farming transition, a process that will likely be country-specific, gradual and complex.

Research on successful tobacco farmer transitions is limited, even in high-income countries such as Canada, Australia and the USA where structural changes to the industry prompted sharp reductions in the number of tobacco farmers. Still, there is a small and growing body of research on transition programs and projects in different contexts from which to draw preliminary lessons regarding the conditions for a successful tobacco-farming transition. These include:

- A national vision for sustainable agriculture
- Removal of public funding for tobacco farming
- Diversification, not substitution
- Access to market infrastructure
- Access to transition funding
- Multi-stakeholder engagement

A national vision for sustainable agriculture

Efforts to foster a transition out of tobacco farming cannot be disconnected from the national system of food production, agricultural land use and rural development within which tobacco is embedded. Fostering new directions in national agricultural policies is consequently central to the transition out of tobacco farming. In this undertaking, ministries of agriculture, rural development, the environment, labor and finance take center stage.

Ministries of health, which typically have the mandate for implementation of the FCTC, can play a supportive role in this process by encouraging other ministries to develop agricultural policies and programs for tobacco-growing regions that respond to a national vision for sustainable agriculture.

Brazil's experiment with a national program for diversification in tobacco-growing areas, while flawed in its implementation, sets a new direction for tobacco farmers based on national policies and programs supporting family agriculture (Almeida, this volume). Principles of environmental sustainability, food security, diversity in land use, participation and partnership are built into the tobacco-farming transition program, drawing on existing public policies aimed at addressing structural problems in the national food system (Medaets et al. 2003). The program provides family farms, including tobacco farmers transitioning to other crops, with access to financial and technical assistance as well as institutional food markets such as school feeding programs. These offer, to some extent, an alternative to the credit system and market guarantee of contract farming and allow tobacco farmers the possibility of breaking free from debt dependency. Importantly, the program fits within a national vision that places the eradication of hunger at the highest level of priority, thereby justifying public investments in land-use changes that promote economically sustainable alternatives to tobacco.

While promising, Brazil's program for diversification in tobacco-growing areas also offers a cautionary tale regarding the limitations of policy change in an environment where the tobacco industry continues to exercise considerable political influence (Almeida, this volume). The program is complicated and has been plagued by numerous bureaucratic failings and chronic underfunding. It was also constrained during negotiations in the Senate to an interpretation that did not allow the program to restrict where tobacco can be cultivated. As a result, contract tobacco farming may simply shift from current areas to new areas where family farms are even poorer and more vulnerable to the pressure tactics of an industry seeking a ready supply of raw materials. National agricultural policies that do not consider where tobacco can be grown are likely to remain only a partial and temporary success.

In low-income countries where food insecurity is a significant issue, agricultural policies that emphasize food production can provide a policy framework for encouraging tobacco transitions. The Ministry of Lands for Bangladesh is considering limiting the expansion of tobacco farming into areas where food crops currently dominate, thereby keeping agricultural lands focused on food production and national food security objectives. An amendment to the Smoking and Tobacco Products Usage (Control) Bill being considered by the government of Bangladesh would also remove all public incentives to tobacco production in food-producing areas. These initiatives put government agencies in Bangladesh in a position to plan and promote a food security approach to the tobacco transition. As shown by Akhter et al. (this volume), the potential of tobacco-growing regions in Bangladesh to address the national deficit in pulses and oilseeds is significant, both locally and regionally. Given that 40 percent of the nation's working population are in non-farming activities that contribute more than 80 percent of the country's export earnings, there is every reason for the government to make the most rational use of scarce agricultural land needed to feed the population.

Research on farmer transitions in the USA and Canada suggests that the model of agriculture guiding the shift out of tobacco matters, not only in terms of the profitability of farming but also community employment effects and benefits to the society as a whole. Transition programs in former tobacco-dependent regions in Kentucky, North Carolina and Maryland funded from resources established through a massive legal settlement with the U.S. tobacco industry resulted mainly in farm consolidation (Capehart 2004). Larger farms and more mechanized farms had the economies of scale needed to produce commodities already known in the region (Russo 2012). There was very little innovation and the total number of land-owning farmers dropped dramatically. Similarly, many tobacco farmers in southern Ontario shifted into ready-made and large-scale industrial cropping systems such as corn and soybean production for ethanol. These systems made little use of the skills farmers had acquired as successful tobacco farmers. Furthermore, replacing a high-value specialized crop that requires local labor (tobacco) with a low-value commodity that is mechanized and capital intensive (corn and soybeans) did little to stimulate employment and other economic spinoffs in farming communities. An interview by the authors with a former tobacco farmer in southern Ontario pointed out that the conversion of many small tobacco farms into a large corn and soybean operation of 1,200 acres displaced dozens of families. He lamented that the entire output of the former tobacco lands was diverted to an inedible product, even though the region has great potential as a food basket for nearby major urban centers.

The experience of the tobacco transition in North America suggests that alternatives to tobacco farming framed only in terms of the international competitiveness of agricultural commodities, as suggested by Keyser (2007), would most likely push smallholder farmers out of farming altogether and further exacerbate a rural exodus already affecting many LMICs. Moreover, the single criterion of international competitiveness reduces agricultural policy to trade policy rather than support for the more complex function agriculture plays in most low- and middle-income countries. A national vision for sustainable agriculture based on the potential of smallholders to meet important and urgent national priorities such as food security and local economic development provides a more promising framework for an alternative future for tobacco-growing regions.

Removal of public funding for tobacco farming

Public investment, even when limited, is an important factor influencing profitability and economic viability for any crop. Recent efforts by the Bangladesh Bank to eliminate loans for the purpose of tobacco farming is a positive step to remove government subsidies that prop up tobacco's profitability (Akhter et al., this volume). Similar steps in Brazil by the central bank (Bancen) are gradually removing subsidies to the tobacco crop while continuing to provide tobacco farmers with access to credit for other crops and land uses (Almeida, this volume). The Brazilian experience also highlights the importance of careful planning and consultation when removing subsidies accessed by tobacco farmers. An overly ambitious timeline and maverick approach to decision-making by the bank almost halted the reform and alienated tobacco farmers when they could have been active supporters of the policy change.

Lebanon represents one of the more complex examples of state intervention in the tobacco sector, involving sizable subsidies to tobacco farmers, comprehensive market supports and direct state investment in promoting the tobacco trade. As shown by Hamade (this volume), these arrangements are determined by political factors, with little concern for the long-term development of rural areas, farmer livelihoods or sustainable food and agriculture at a national level.

Countries where some form of supply management is still in place can also learn from the mistakes of tobacco reform in North America. The dismantling of a longstanding supply-management system was the focus of structural changes in the tobacco industry in the USA and Canada between 2004 and 2008. This typically involved a buy-out of tobacco marketing quotas and an end to price supports that had made tobacco one of the most profitable crops in North America until tobacco companies began to shift their operations to lower-cost production centers in LMICs. Governments also made public investments in research on alternative crops and provided marketing assistance (Russo 2012; Griffith 2009). Despite these financial and technical supports, the transition met with bitter resistance from tobacco farmers who resented this loss of privilege. Benson (2011, 4) argues that "tobacco industry propaganda […] goaded the grower ranks into a collective feeling of being conspired against, even though there isn't any evidence of a concerted attack on tobacco livelihoods waged by the government and public health groups."

As noted above, simply dismantling the supply-management system did not result in an overall decline in tobacco production but rather a reduction in the number of farms. Direct subsidies to the remaining farms by the U.S. government continued unabated, totaling USD 1,519,000,000 from 1995 to 2012.[3] These subsidies, more than any inherent efficiency in the U.S. tobacco industry, has kept tobacco farming relatively profitable for the larger mechanized farms. Griffith (2009) argues that tobacco corporations actually welcomed the dismantling of the supply-management system because it made it easier to convert the remaining tobacco farmers into quasi-wage workers harnessed to the contract farming system. This North American experience suggests that the dismantling of market and price support programs by governments must go hand in hand with the removal of other direct subsidies to the production of tobacco. Otherwise, public resources will continue to be siphoned off by tobacco corporations seeking to reduce their costs and increase their power in the agricultural economy.

Direct participation of governments in co-ownership of the tobacco industry is perhaps the most troubling ongoing misuse of public funding in the sector. In Southeast Asia (Barraclough and Morrow 2010), government participation in the tobacco industry includes ownership of tobacco factories (Myanmar), tobacco marketing monopolies (Vietnam, Thailand) and direct capital investment in private tobacco companies (Laos, Singapore, Indonesia, Malaysia). State ownership of shares in tobacco manufacturing and in the tobacco trade is also common in Africa (Kenya, Malawi, Ethiopia), the Middle East (Lebanon), Latin America (Argentina, Brazil) and countries of the former Soviet Union. China is perhaps the most glaring example – a single agency (State Tobacco Monopoly Administration) is responsible for both implementing effective tobacco-control policies and ensuring the effective development of the commercial production of tobacco (Wan et al. 2012). This widespread situation of government co-ownership of the industry is a major contradiction to Article 5.3 of the FCTC, which aims to protect public health policies from tobacco industry influence. It also directly compromises the regulatory role of governments and perpetuates imbalances in the leaf marketing chain that favor returns to tobacco corporations and investment capital rather than income to farmers. As argued by Barraclough and Morrow (2010, 49), resolving this fundamental contradiction requires that governments "commit themselves to disinvest from the tobacco industry and to cease the promotion of tobacco." This is a fundamental element in any tobacco-farming transition, without which the promotion of alternatives will always be at an economic and political disadvantage.

---

3   Environmental Working Group. 2013 Farm Subsidy Database. Online: http://farm.
     ewg.org/top_recips.php?fips=00000&progcode=tobacco (accessed 3 January 2014).

Diversification, not substitution

Most research on economically sustainable alternatives to tobacco has focused on the identification of an individual crop to substitute directly for tobacco, in the same field at the same time. For example, a project in China focused on the production of high-value vegetables, mushrooms and arrowhead that have similar or better net gains for farmers (Li et al. 2012). In Aytaroun (Lebanon) local activists are experimenting with a popular kind of local oregano *(zaatar)* as a substitute for the tobacco crop (Hamade, this volume). For some farmers in Kentucky (USA), sweet potatoes successfully substituted for tobacco while in parts of Ontario (Canada) ginseng is grown on former tobacco lands.

While substitute crops are likely to be essential components of economically sustainable alternatives to tobacco growing, single crops are not magic bullets and are unlikely to be the answer for smallholders in most regions where tobacco is currently grown. The reasons for this are twofold. First, it is technically difficult and costly to simply insert another monocrop into the same space as tobacco and in the same season without first improving the soil. Tobacco is notoriously hard on the soil, not only depleting nutrients and soil organic matter but also exacerbating weed infestation, soil erosion, soil pollution and salinization (Lecours, this volume). While in some regions farmers have been able to manage tobacco's deleterious effects on the soil through rotations, farmers with small plots, as in Malawi and Bangladesh, have had little choice but to mine the soil to exhaustion and intensify the use of costly agrochemicals.

Second, a focus on the direct substitution of another crop limits the range and diversity of cropping systems and livelihoods options farmers can consider. Tobacco is a long duration crop, with a period of nursery development at the beginning, a single harvest at the end of the growing period and a period of transformation (curing) before the crop can be sold. In many environments this cycle occupies part but not all of the potential growing season, thereby shutting out the possibility of crops with different planting and harvesting times, linkages between crops and livestock, better water use, a distributed cash flow, reduced household food expenditures and incremental improvements to soil conditions, among other possibilities. A focus on crop and livelihood diversification rather than a single substitute crop offers more scope for innovation and the use of all resources available, including minor seasons, marginal soils, value-added production and off-farm employment. This approach is in keeping with methodological guidelines developed by Brazil as a contribution to work on policy options and recommendations concerning Articles 17 and 18 of the FCTC (Schneider et al. 2012).

Experience from Bangladesh illustrates what can be done by creating transition strategies based on diversification rather than simple substitution

(Akhter et al., this volume). Farmers first introduced a mix of short cycle food crops including legumes to help improve soil conditions for later crops while also producing a modest food and cash income stream. They then inserted a variety of food, fiber and fodder crops into their fields with shorter and longer cycles, including a few high-value crops such as jute, potatoes and peanuts. Mixed cropping and crop diversity created positive interactions with livestock (the production of fodder for sale or for their own animals), reduced household expenditures on food and created enough cash income to break their dependency on bridging loans and advances on tobacco harvests. The cropping system as a whole provided a positive net benefit, taking into account factors beyond the yield of a single crop grown in a monoculture. It also drew on farmer knowledge of crop diversity and fostered thinking "outside the box" of a single alternative crop.

Research in Kenya also took a farming systems approach to the transition out of tobacco, albeit with a single crop (bamboo) playing a leading role. Farmers there had larger tracts of land than in Bangladesh and were already growing a number of crops other than tobacco. Bamboo was initially introduced into under-utilized areas on their farms such as around the homestead, on river banks or around the perimeter of a field. It was also introduced into less productive tobacco fields and intercropped with seasonal crops such as kale and other vegetables with ready markets or needed for home consumption. The amount of land shifted out of tobacco at any one time also proceeded slowly, allowing farmers to maintain an income from tobacco while they waited for bamboo to reach maturity (Kibwage et al., this volume). Gradually, farmers extended bamboo to more and more of their farm, eventually resulting in a shift to an agroforestry system with a mix of trees (bamboo) and food and cash crops.

The experience in Bangladesh, Kenya and Brazil reported in this book, and findings elsewhere (Chavez et al. 2012; Geist et al. 2009), suggest that a focus on cropping and farming system change, diversification and on-farm innovation may offer greater scope for successful transitions, at lower cost, than top-down, large-scale tobacco conversion projects involving a single crop and homogenous production systems.

Access to market infrastructure

Various studies show that the market guarantee provided through contract tobacco farming is compelling for farmers with limited access to other markets and market infrastructure. Studies in Brazil (Vargas and Campos 2005), Southeast Asia (SEATCA 2008) and Africa (Kagaruki 2010; Otañez et al. 2006) all conclude that the most important reasons given by farmers for growing tobacco is that contracts provide working capital at the beginning of the season and an assured market for the harvest. Similar reasoning is present in Kenya, Brazil and Bangladesh, where contract farming also dominates the industry (see case studies, this volume).

The challenge for a tobacco-farming transition is to overcome the effects of many years of decline in traditional markets and limited government investment in market infrastructure. Even in middle-income countries with relatively large agricultural economies such as Brazil, tobacco farming often exists within a localized market desert where whole communities have become dependent on a single commodity and a single buyer. Creating conditions for a tobacco-farming transition will consequently need to find low-cost and sustainable means to enhance access to markets and market infrastructure supportive of a wider variety of agricultural products.

The case studies presented in this book point to a range of strategies in response to the challenges of market access and infrastructure development. The Brazil program on tobacco diversification facilitates smallholder access to institutional food buyers through existing national programs such as the national food acquisition program and the school feeding program (Almeida, this volume). These policies are designed to provide a social safety net for the food insecure while also stimulating food production by family farms. While the programs are open to all smallholder farmers, the tobacco transition program provides targeted support to ensure that tobacco farmers know about the program and have access to its facilities. So far, the institutional buyers have not reached the minimum targets set for purchases from family farms in tobacco-growing regions because food production there does not yet meet the required criteria in terms of volume and quality. However, early findings suggest that the program is stimulating food production and that tobacco farmers are beginning to participate more actively in the program (Almeida, this volume).

Kibwage et al. (this volume) found that an entirely new market was needed through which to channel bamboo production used to substitute for tobacco production. Market feasibility studies identified two key barriers. First, a 1986 Presidential Order banning the harvesting of bamboo from national forests made it difficult to distinguish between illegally harvested bamboo from legally grown bamboo. This was overcome by creating cooperatives of bamboo producers authorized by the government to sell their products. Second, low-cost plastic and metal products compete directly with products also manufactured from bamboo. Alternative bamboo products were needed that added value to the raw material produced by farmers and that farmers could produce themselves. This was accomplished by drawing on the skills and existing infrastructure of tobacco farmers. Curing barns used to cure tobacco were converted at little cost into spaces for curing bamboo poles with a ready market as scaffolding material in the construction industry. Nurseries used to grow tobacco seedlings, and the knowledge associated with running a nursery, enabled other tobacco farmers to establish nurseries for bamboo seedlings and bamboo ornamental plants that could also be marketed directly or through farmers' cooperatives. Government representatives of the National Environment Management Agency bought large numbers of bamboo seedlings for reforestation from organized farmers, providing an early market for one of the most profitable bamboo products. Finally, craft skills existing in the villages were harnessed to create high-value products from bamboo (furniture, mats, etc.) with established urban markets. These provided high rates of return, and strong incentives for farmers to convert tobacco land and under-utilized lands into bamboo production.

In Bangladesh (Akhter et al., this volume), market supports focused on the infrastructure for aggregation, storage and transportation of agricultural products. First, external (project) resources were used to rent cold storage facilities farmers could use to aggregate and store their product during periods of oversupply and low prices. The economy of scale created by aggregated products and collective organization in turn reduced individual costs for transportation and marketing of goods in local and regional wholesale markets. The farmer organization Nayakrishi Andolon also extended to former tobacco farmers the market niche it had developed for higher-value organically grown fruits and vegetables. The reputation and marketing infrastructure of the Nayakrishi Andolon provided former tobacco farmers with access to more distant urban markets with premium prices. Hamade (this volume) suggests that the state tobacco monopoly in Lebanon (the Régie), under the direction of the Ministry of Finance, could play a similar role by providing tobacco farmers in transition with access to its regional infrastructure (storage facilities and transportation services).

Experience elsewhere suggests that technical research and rural extension can also contribute to addressing key constraints in value chains and market infrastructure. In Malawi, research and training on ways to manage the aflatoxin problem in groundnuts has begun to reopen export markets for Malawian groundnuts. The crop, well adapted to the degraded soils of tobacco lands, now offers tobacco farmers with an option that had been lost to them due to contamination. The National Smallholder Farmers' Association (NASFAM) – a membership organization of more than 120,000 small-scale farmers – has picked up this research and now offers its members an aflatoxin screening service, help with packaging and transportation of groundnuts and access to a fair trade market agreement negotiated with international buyers. As of January 2014, the association stated on its website that it now focuses on promoting "diversification away from dependency on maize and tobacco, and thus supports production and marketing of crops such as groundnuts, chili, rice, soya, beans, sunflower and others as appropriate to market demand."

A local food, direct-marketing strategy was the focus for marketing assistance provided by state agriculture departments, universities and economic development agencies in a number of former tobacco-dependent regions in the southern USA. While it was successful at increasing exponentially the amount of food produced and sold by local farms, only some farms made the shift to mixed farming systems, market gardens or niche food products. Most farms ultimately converted from tobacco to one or two commodities for which there were already established national markets and industrial scale production systems (poultry, horses, pigs). The exception was southern Maryland where farmers had good access to urban markets and a strong tradition of food-based agriculture among the majority Amish religious community. The local food, direct-marketing strategy, and in particular "buy local" marketing campaigns in nearby urban centers, built on these favorable conditions. In the absence of a single alternative to tobacco, farmers were forced to be creative and to make use of the agricultural knowledge and assets they had to develop new products and markets (Russo 2012).

Access to financing

Adjustment to new cropping systems and product markets takes time, creativity and resources. Keyser (2007) suggests that transitions can be financed from within the farm system by using the income from tobacco to invest in new activities. While this is possible in some cases, low net profits, long-term debt and a lack of infrastructure for accumulating savings are significant barriers for smallholder tobacco farmers in many low- and middle-income countries. Even in the highly subsidized tobacco-farming economy of Lebanon the net return on tobacco farming is very low (Hamade, this volume). For these farmers, and most other smallholders currently dependent on tobacco production as their sole or primary source of income, some form of financial assistance for transition is necessary. This is especially the case for smallholders with land degraded by tobacco production because even with mixed cropping, composting and crop rotations the recovery of land takes time. Akhter at al. (this volume) suggest that very degraded tobacco soils in Bangladesh can take two to three years to recover their natural potential. These circumstances justify investment by governments in start-up capital and support to cover the incremental costs of the transition period.

Countries with micro-credit and other loan programs for directing investment in small-scale agricultural enterprises could be adapted to meet the transition costs of tobacco farmers. Brazil's tobacco diversification program anticipates this need by linking the program to national credit schemes for family agriculture, including financing for planning crop transitions (Almeida, this volume). Fiscal reforms in which additional taxes are collected on tobacco consumption are one way to contribute to these financing mechanisms.

In countries with supply management programs such as Lebanon, tobacco consumption taxes could also be used to buy out licenses and quotas, much as governments in the USA, Canada and Australia have done in recent years. A successful smallholder transition in southern Ontario, for example, relied on a combination of advanced planning of land-use and farming system changes (by diverting some farmland to pasture prior to shifting out of tobacco) and the careful use of the tobacco quota buy-out funds to develop new, farm-based market infrastructure for organic beef production.[1] The risks of poorly conceived and implemented transition financing are worth noting, however, and are no more evident than in the Canadian case. In southern Ontario, where much of Canada's tobacco farming was and still is concentrated, tobacco production remains at about 30 million kg annually, less than at its peak in the 1970s and 1980s but higher than when the transition program began in 2008.[5] Recent reporting on the buy-out program by the Auditor General of Canada found that almost half of the farmers receiving funding were not active tobacco farmers at the time and many others simply transferred their land and tobacco equipment to close relatives and kept on growing tobacco under a new licensing system (Mann 2011). The Canadian Taxpayers Association website, accessed 23 January 2014, lists this program of Agriculture Canada as the winner of the "Teddy Award" for the worst example of federal government waste in 2012.

Multi-stakeholder engagement

Changing how tobacco farming fits within the broader system of agricultural land use, the removal of public funding for tobacco farming, diversification, facilitating access to market infrastructure and bridging the farm-level costs of transition are complex undertakings. Tobacco-farming transitions will consequently require committed multi-stakeholder engagement, coordination and political resolve. This will not be easy given the high level of influence of the tobacco industry on governments and government policy, especially in LMICs where tobacco farming is prominent in the agricultural economy.

Jones et al. (2008) suggests that building alliances between the tobacco-control movement and tobacco farmers has the potential to provide a

---

4   Personal communication, Bryan Gilvesy (rancher and former–tobacco grower), 5 January 2014.

5   Personal communication, Art Lawson, South Central Ontario Region Economic Development Corporation, 5 January 2014.

counterbalance to the current influence of the tobacco industry on government policy, and create buy-in from tobacco farmers. Effective engagement of tobacco farmers in Bangladesh led to the development of an action-oriented alliance between urban-based anti-smoking groups and a high profile national farmers' movement (the Nayakrishi Andolon). While non-governmental organizations, health agencies and advocates in Bangladesh had previously focused exclusively on the negative effects of smoking on the health of the general population, new awareness and knowledge about the harmful impact of tobacco farming on the health of tobacco workers, the environment and national food security objectives broadened the discourse. Tobacco farmers in transition participating in World No Tobacco Day demonstrations in urban centers also captured the interest of the national media and helped to raise public understanding of the full range of harms from tobacco. These initiatives built bridges between the Ministry of Health, responsible for implementation of the FCTC, and other government agencies with mandates relevant to the development of tobacco transition strategies in tobacco-dependent regions. In effect, farmer engagement helped to inform and broaden the tobacco-control debate by making the diversification issue not only a question of agricultural policy but a public health issue as well (Smith et al. 2000).

The Kenya experience also points to the relevance of action-oriented farmer engagement to creating the conditions for tobacco transition strategies. Tobacco farmers were first consulted regarding the broader livelihood problems they were encountering and whether or not and how they wanted to engage with researchers in experiments with alternatives to tobacco. Successful farmer mobilization around these experiments attracted the support of local politicians and leaders of churches and non-governmental organizations, who in turn generated political interest in project results at the highest levels of government. This facilitated policy changes (legal harvesting of bamboo) needed to support bamboo production and created government support and direct assistance from government agencies for the creation of producer cooperatives. Farmer organization proceeded more quickly and acquired a level of legitimacy it would not have gained otherwise (Kibwage, this volume).

Almeida's study (this volume) shows that the national diversification program in Brazil offers tobacco farmers a powerful platform to negotiate assistance with key transition strategies, drawing on a wide range of existing national agricultural policies and programs. The emphasis on diversification as a means to enhance farmer income and sustainability may also serve to inform discussion regarding future directions for family farming across the whole agricultural sector, not just in tobacco-growing regions. Thus, the potential contribution of engagement with tobacco farmers in Brazil is multi-directional because it serves to situate tobacco control in broader debates regarding social justice for family farming, environmental sustainability and public health.

Bottom-up and top-down strategies are needed both to strengthen farmer organization where none exists and provide organized farmers with opportunities to join in the fight against the tobacco epidemic and creatively address the tobacco dependence of farmers and farm communities. In this mix, multi-stakeholder engagement, dialogue and action-oriented research have vital roles to play in creating buy-in from the key actors and support for tobacco-farming transitions.

Conclusion

Controlling a powerful and well-resourced industry that produces an inherently harmful product has been an enormous challenge ever since the connection between tobacco use and human disease was established definitively in the 1950s. During the early years, tobacco companies in the USA made effective use of their relationship with tobacco farmers to undermine tobacco control through what a representative of R. J. Reynolds Tobacco Company called "complete industry unity" (quoted in Jones et al. 2008). The industry–farmer alliance, while not without its difficulties, was a powerful and effective political lobby for decades.

The harsh realities of tobacco farming in the current era, particularly in low- and middle-income countries where production is now concentrated, have made this alliance more difficult to sustain. The industry lost much of the support it had from tobacco farmers in developed countries like the USA and Canada who were increasingly disheartened by changes in the industry. Prices declined when global production sharply turned to the Global South and when new cigarette manufacturing technology reduced the need for high quality tobacco leaf (Chaaban, this volume). While tobacco corporations have attempted to recreate the idea of an alliance with developing country tobacco farmers, in reality its relationships are typically with farmer front groups of their own making (Assunta 2012) and government agencies with vested interests in keeping the industry going. These appearances of partnership, as misleading as they are, still carry weight in the policy making process and confuse what are straightforward observations regarding tobacco control and tobacco farmers.

The observations are fourfold. First, today's tobacco farmers have nothing to fear from tobacco control in their own countries as a source of instability and price volatility affecting their bottom line. This is true in virtually every country where tobacco is primarily an export crop, as it is in Africa, Latin America and much of Asia. Only in China and India, where a large proportion of production is for domestic consumption, is tobacco farmer income potentially affected by declining domestic rates of tobacco use. Even in these countries the export of tobacco leaf and manufactured cigarettes also mitigates against the risk of declining domestic demand. Clearly, domestic tobacco-control policies can be pursued aggressively by all tobacco-exporting nations, knowing that global demand for tobacco leaf is likely to continue at current or higher levels for some time. This shelters the livelihoods of the current generation of tobacco farmers from domestic policy change.

Second, there are many good reasons why governments should create the conditions for a transition out of tobacco farming now, rather than delaying the challenge indefinitely. Tobacco industry practices such as child labor, underpricing and under-grading of tobacco, overpricing of input packages and the sponsorship of long-term debt through contract farming are significant factors negatively affecting farmer livelihoods. While historically tobacco farming was profitable under some conditions, today in many LMICs tobacco farming produces very low or negative net returns to land and labor. The mining of soils and forests to fuel the tobacco industry also adds to the downward spiral of exploitation, vulnerability and dependence experienced by tobacco farmers. Many tobacco farmers would like to move out of tobacco growing, but they feel they cannot.

Third, evidence from a variety of settings shows that many other crops, crop combinations, farming systems and livelihood strategies offer better opportunities for farmers, if they can break free from their immediate dependency on the tobacco market guarantee and loans provided through contract farming. This is not easy, and for most smallholders in LMICs the transition cannot be financed from within the farm itself. Access to public resources is necessary, especially in the form of access to credit, technical assistance and the development of local and regional market infrastructure. Divestment of public funds from tobacco enterprises and the removal of all public funding to the tobacco industry is also a vital part of creating the conditions for the development of economically sustainable alternatives to tobacco farming. In this process, tobacco farmers should not be blamed for growing tobacco or treated as passive victims of an all-powerful industry. Rather, the task is to engage the tobacco-farming community and other stakeholders in the development of new options and better rural livelihoods.

Finally, the tobacco transition must consider how tobacco farming fits in the broader system of rural development, agricultural land use and agricultural policy reform and programs. Broadly speaking, this means investing in a multi-faceted development strategy that can begin to overcome the effects on smallholder agriculture of decades of declining government support. In this discussion, Article 17 of the FCTC provides a constructive and modest space for the tobacco-control community and health ministries to play a supportive role to ministries of agriculture, the environment, labor, finance and rural development. These actors can take center stage as advocates for smallholder farmers and the pursuit of broader social goals such as food security and the diversification of rural livelihoods. Only then can governments set themselves on a track towards a fair and orderly transition consistent with the goals of tobacco control and the spirit of the FCTC.

References

Akhter et al., this volume.

Almeida, this volume.

Amigó, M. F. 2010. "Small Bodies, Large Contribution: Children's Work in the Tobacco Plantations of Lombok, Indonesia." *Asia Pacific Journal of Anthropology* 11: 34–51.

Assunta, M. 2012. "Tobacco Industry's ITGA Fights FCTC Implementation in the Uruguay Negotiations." *Tobacco Control*. DOI:10.1136/tobaccocontrol-2011-050222. 6 pp. (accessed 7 December 2013).

Barraclough, S. and M. Morrow. 2010. "The Political Economy of Tobacco and Poverty Alleviation in Southeast Asia: Contradictions in the Role of the State." *Global Health Promotion* Supplement (1): 40–50.

Benson, P. 2011. *Tobacco Capitalism: Growers, Migrant Workers, and the Changing Face of a Global Industry*. Princeton: Princeton University Press.

Callard, C. D. and N. E. Collishaw. 2013. "Supply-Side Options for an Endgame for the Tobacco Industry." *Tobacco Control* 22: 10–13.

Capehart, T. 2004. "Trends in U.S. Tobacco Farming." Electronic Outlook Report from the Economic Research Service, TBS-257-02. United States Department of Agriculture. 7 pp.

Centers for Disease Control and Prevention. 2007. *Best Practices for Comprehensive Tobacco Control Programs–2007*. Atlanta, GA: U.S. Department of Health and Human Services.

Chaaban, this volume.

Chavez, M. D., P. B. M. Berentsen and A. G. J. M. Oude Lansink. 2012. "Assessment of Criteria and Farming Activities for Tobacco Diversification Using the Analytical Hierarchical Process (AHP) Technique." *Agricultural Systems* 111: 53–62.

Drope, J. 2011. "Conclusion: Tobacco Control in Africa – People, Politics and Policies." In *Tobacco Control in Africa – People, Politics and Policies*, edited by J. Drope, 283–94. London: Anthem Press.

Eriksen, M., J. Mackay and H. Ross. 2012. *The Tobacco Atlas, Fourth Ed*. New York, NY: World Lung Foundation.

ETC Group. 2013. "Putting the Cartel before the Horse …and Farm, Seeds, Soil, Peasants, etc." *Communiqué* No. 111, September 2013, 40 pp.

Geist, H. J., K. Chang, V. Etges and J. M. Abdallah. 2009. "Tobacco Growers at the Crossroads: Towards a Comparison of Diversification and Ecosystem Impacts." *Land Use Policy*. 26: 1066–79.

George, S. 1988. *A Fate Worse Than Debt: The World Financial Crisis and the Poor*. New York: Grove/Atlantic Press.

Glantz, S. and M. Gonzalez. 2011. "Effective Tobacco Control Is Key to Rapid Progress in Reduction of Non-Communicable Diseases." Lancet: DOI:10.1016/S0140-6736(11)60615-6.

Griffith, D. 2009. "The Moral Economy of Tobacco." *American Anthropologist* 111(4): 432–42.

Hamade, this volume.

Hipple, B., J. Lando, J. Klein and J. Winickoff. 2011. "Global Teens and Tobacco: A Review of the Globalization of the Tobacco Epidemic." *Current Problems in Pediatric and Adolescent Health Care*. 41: 216–30.

Jha, P. and F. J. Chaloupka. 1999. *Curbing the Epidemic: Governments and the Economics of Tobacco Control*. Washington: The World Bank.

Jones, A. S., W. D. Austin, R. H. Beach and D. G. Altman. 2008. "Tobacco Farmers and Tobacco Manufacturers: Implications for Tobacco Control in Tobacco-Growing Developing Countries." *Journal of Public Health Policy* 29: 406–23.

Kagaruki, L. 2010. "Community-Based Advocacy Opportunities for Tobacco Control: Experience from Tanzania." *Global Health Promotion* Supp (2): 41–44

Keyser, J. C. 2007. "Crop Substitution and Alternative Crops for Tobacco." Study conducted as a technical document for the first meeting of the ad hoc study group on alternative crops established by the conference of the parties to the WHO framework convention on tobacco control. Lusaka, 54 pp.

Kibwage et al., this volume.

Lecours, this volume.

Lee, S., P. M. Ling and S. A. Glantz. 2012. "The Vector of the Tobacco Epidemic: Tobacco Industry Practices in Low and Middle-Income Countries." *Cancer Causes Control* 23: 117–29.

Li, V. C., Q. Wang, N. Xia, S. Tan and C. Wang. 2012. "Tobacco Crop Substitution: Pilot Effort in China." *American Journal of Public Health* 102(9): 1660–63.

Malone, R. E., Q. Grundy and L. A. Bero. 2012. "Tobacco Industry Denormalisation as a Tobacco Control Intervention: A Review." *Tobacco Control* 21: 162–70.

Mann, S. 2011. "Audit Takes Aim at Tobacco Buyout." *Better Farming: Ontario's Online Community of Professional Farmers*, 24 November 2011. Online: http://www.betterfarming.com/online-news/audit-takes-aim-tobacco-buyout-4672 (accessed 23 January 2014).

Medaets, J. P., K. Pettan and M. Takagi. 2003. "Family Farming and Food Security in Brazil. Ministry of Agrarian Development and Extraordinary Ministry for Food Security." Online: http://www.oecd.org/tad/25836756.pdf (accessed 6 January 2014).

Nakkash, R., J. Khalil and R. Afifi. 2011. "The Rise of Narghile (Shisha, Hookah) Waterpipe Tobacco Smoking: A Qualitative Study of Perceptions of Smokers and Non Smokers." *BMC Public Health* 11: 315 DOI:10.1186/1471-2458-11-315.

Oluwafemi Mimiko, N. 2012. *Globalization: The Politics of Global Economic Relations and International Business*. Durham, N.C.: Carolina Academic.

Otañez and Graen, this volume.

Otañez, M., G. Glantz and H. Mamudu. 2009. "Tobacco Companies' Use of Developing Countries' Economic Reliance on Tobacco to Lobby against Global Tobacco Control: The Case of Malawi." *American Journal of Public Health* 99(10): 1759–71.

Otañez, M. G., M. E. Muggli, R. D. Hurt and S. A. Glantz. 2006. "Eliminating Child Labor in Malawi: A British American Tobacco Corporate Responsibility Project to Sidestep Tobacco Labor Exploitation." *Tobacco Control* 15: 224–30.

Roy, A., D. Efroymson, L. Jones, S. Ahmed, I. Arafat, R. Sarker and S. Fitzgerald. 2012. "Gainfully Employed? An Inquiry into Bidi-Dependent Livelihoods in Bangladesh." *Tobacco Control* 21(3): 313–17.

Russo, R. A. 2012. "Local Food Initiatives in Tobacco Transitions of the Southeastern United States." *Southeastern Geographer* 52(1): 55–69.

Schneider, S., P. Waquil, L. Xavier, M. Conterato, M. Perondi, A. G. Rambo, C. Schneider Rudnick, T. Dias Freitas and K. F. de Avila. 2012. "Methodological Guidelines for the Analysis of Tobacco Growers Livelihoods Diversification." Draft Version. Online: http://www1.inca.gov.br/tabagismo/publicacoes/methodological_guidelines_brazil. pdf (accessed 23 January 2014).

SEATCA. 2008. "Cycle of Poverty in Tobacco Farming: Tobacco Cultivation in Southeast Asia." Bangkok: Southeast Asia Tobacco Control Alliance. 13 pp.

Smith, M. H., D. G. Altman and B. Strunk. 2000. "Readiness to Change: Newspaper coverage of tobacco farming and diversification." *Health Education and Behavior* 27(6): 708–23.

Stein, H. 2010. "World Bank Agricultural Policies, Poverty and Income Inequality in Sub-Saharan Africa." *Cambridge Journal of Regions, Economy and Society* August:1–12.

Tobin, R. and W. Knausenberger. 1998. "Dilemmas of Development: Burley Tobacco, the Environment and Economic Growth in Malawi." *Journal of Southern African Studies* 24(2): 405–24.

Townsend, J. L., P. Roderick and J. Cooper. 1994. "Cigarette Smoking by Socioeconomic Group, Sex, and Age: Effects of Price, Income, and Health Publicity." *British Medical Journal* 309: 923–26.

Vargas, M. A. and R. R. Campos. 2005. "Crop Substitution and Diversification Strategies: Empirical Evidence from Selected Brazilian Municipalities." Discussion Paper, Economics of Tobacco Control, No. 28 published by The World Bank, Washington, 50 pp.

Wan, X., S. Ma, J. Hoek, J. Yang, L. Wu, J. Zhou and G. Yang. 2012. "Conflict of Interest and FCTC Implementation in China." *Tobacco Control* 21: 412–15.

Warner, K. 2000. "The Economics of Tobacco: Myths and Realities." *Tobacco Control* 9: 78–89.

Warner, K., G. Fulton, P. Nicolas and D. R. Grimes. 1996. "Employment Implications of Declining Tobacco Product Sales for the Regional Economies of the United States."

Annex

A POLICY BRIEF ON TOBACCO CONTROL AND TOBACCO FARMING

Context and Importance of the Problem

Despite the unarguable merits of tobacco control, implementation of the Framework Convention on Tobacco Control (FCTC) is only just beginning in many countries. The slow pace of implementation costs countless lives and imposes economic hardship on governments faced with rising healthcare costs and lost opportunities to invest in sustainable development. There is no time or need to delay.

The influence of the tobacco industry lobby on governments and policy makers has been a significant factor behind delays in implementation of the FCTC. Strategies used for decades by the tobacco industry to dilute, delay and defeat tobacco control in high-income countries are now being redeployed successfully in low- and middle-income countries (LMICs) where most of the world's tobacco is now grown and where growth in tobacco consumption is greatest. The thrust of their strategy is to create a fear of tobacco control among policy makers where there should be none. Despite evidence to the contrary, industry representatives claim that:

- Measures to control tobacco use will provoke a livelihood crisis among tobacco farmers and workers in the industry;
- Tobacco farmers are currently relatively prosperous and tobacco farming poses no significant risks that cannot be mitigated;
- There are no economically sustainable alternatives to tobacco farming for smallholder farmers, particularly in low- and middle-income countries.

These myths have created among some governments and policy makers the false perception that alternative livelihoods must be in place *before* any further action is taken to implement tobacco-control measures in their country. Nothing could be further from the truth. Policies for tobacco control can proceed without affecting the livelihood of tobacco farmers. The creation of economically sustainable alternatives to tobacco farming can proceed in parallel, in the interest of improving livelihoods and protecting the environment. There are many good reasons why governments should promote both tobacco-control policies and the development of economically sustainable alternatives for tobacco growers, as outlined below.

Why the Industry Claims Are Myths

Industry investment analysts and health organizations alike confirm that despite substantial progress in implementing demand reduction measures in many countries there is no indication of an impending rapid decline in global demand for tobacco leaf. While a number of high-income countries have been successful at reducing consumption, this has been more than offset by the growth of consumption in low- and middle-income countries where population increases and aggressive tobacco marketing have stimulated an expansion of the global market. A recent study suggests that in the most optimistic scenario, the number of smokers worldwide could go down from 794 million in 2010 to 523 million in 2030 (a drop of 34 percent). In the absence of substantial further tobacco-control measures, the number would actually grow (by almost 10 percent) to 872 million.[1] In either case, the global market will remain significant during the lifetime of the current generation of tobacco farmers. Negative impacts of tobacco-control measures on farmer livelihoods could only happen in a closed national economy where all of the tobacco produced is consumed in the country and there are no exports or imports of tobacco – a scenario not seen, nor likely to be seen in any country. In fact, the vast majority of farmers in low- and middle-income countries produce for the global market and are consequently unaffected by demand reduction measures in their own countries.[2] The drivers of demand and prices for tobacco leaf, a central concern of tobacco farmers and policy makers alike, lie elsewhere and are grounded in the business model of the tobacco industry. These include:

- **Lowest labor costs and weakest environmental standards:** Tobacco farming is more labor intensive than most other crops and, in

---

1   Mendez, D., O. Alshanqeety and K. E. Warner. 2013. "The Potential Impact of Smoking Control Policies on Future Global Smoking Trends." *Tobacco Control* 22(1): 46–51.

2   Even in China, where a lot of the lower quality tobacco produced is only consumed locally, and in the absence of more effective tobacco-control policies, the factors that explain the imminent reduction in demand for lower-quality leaf have to do with the increased competition from higher quality tobacco imports, occurring as a result of China joining the World Trade Organization (T. Hu, Z. Mao, H. Jiang, M. Tao, and A. Yurekli. 2007. "The Role of Government in Tobacco Leaf Production in China: National and Local Interventions." *International Journal of Public Policy* 2(3/4): 235–48).

the case of flue-cured tobacco, requires large amounts of fuel wood from forests. The business model of the tobacco industry has been to seek out the lowest farm labor costs and weakest environmental standards it can find in environments suitable for tobacco cultivation. This has prompted a rapid shift in tobacco production from high-income countries to low- and middle-income countries with cheap labor and weak or poorly enforced environmental policies and programs, especially with respect to forests and soil resources. Today, 90 percent of commercial tobacco leaf is grown in LMICs, compared to much lower rates only a few decades ago.

- **Vertical integration:** The tobacco industry is a vertically integrated supply and marketing chain bound through a few common owners of subsidiary companies and contracted farmers. This has created a market with few buyers (a handful of transnational corporations) and many sellers (hundreds of thousands of largely unorganized tobacco farmers). Under these monopolistic conditions, the buyers can play sellers in different settings, regions and countries off against one another, to lower their costs. Through contract farming the tobacco companies can also dictate what, how and when tobacco leaf is supplied, and pass off to farmers the risks of over-production in one place and losses due to climate variability and environmental degradation in another. Vertically integrated companies can also shift costs and income between company branches and country subsidiaries in order to avoid corporate taxes and maximize profits at selected stages in the value chain.

- **Declining public investment in smallholder agriculture:** Decades of structural adjustment policies and unequal trade regimes have undermined national and international investment in diversified smallholder agriculture. This decline and reorientation of investment shifted production in many low- and middle-income countries to export crops favoring economies of scale (large farms) and the lowest price on the global market for raw materials. The resulting decline in domestic agricultural markets, government extension support and market infrastructure has left smallholders with few meaningful choices, a vacuum the tobacco industry filled through its monopoly practices.

Recognition of the underlying drivers of global demand for tobacco leaf and price volatility affecting today's tobacco farmers helps to put the industry in perspective, and examine the harsh realities of tobacco farming. Tobacco farming as it is practiced in LMICs today does not lift farmers out of poverty. The profitability of tobacco farming quickly evaporates into losses once even the most basic costs such as labor and inputs are factored into the calculation of net gains. Instead, many tobacco farmers are trapped by the accumulation of debt, the exploitation of child labor and the selling of false hopes for seasons to come. Meanwhile, exposure to unique occupational health hazards is borne by the entire family, field workers and neighboring communities. The mining of soils and forests to fuel the tobacco industry adds to exploitation, vulnerability and dependence. Due to these features, the worldwide tobacco pandemic is rightly seen not only as an important health issue but also a development issue holding back the creation of income and savings normally expected from a viable farming operation and general advancement of the rural economy.

The measures required to curb tobacco consumption are relatively well understood. They are based on solid evidence collected over many years and in many different countries regarding, for example, the introduction of higher taxes on tobacco products, legislated health warnings on tobacco packs, bans on advertising, promotion and sponsorship, and the creation of smoke-free environments. The same cannot be said of the information and knowledge available about the conditions for a tobacco-farming transition. While research has shown that there exist a variety of context-specific alternatives to tobacco, less is known about how to scale-up these initiatives in the context of the sector-wide challenges facing smallholder farmers as a whole. This uncertainty increases the complexity of policy interventions, not because there are no alternatives to tobacco farming but rather because the goal is long-term development of a strong and diversified agricultural economy. Pitting tobacco-control policies against farmers and the legitimate search for better livelihoods is actually a false dilemma promoted by the tobacco industry with a single purpose – to undermine the national and international consensus on the urgency of tobacco control as both a public health and a development policy.

Implications and Recommendations

The issues that should form the core of government policies for tobacco control and tobacco farming revolve around the need to urgently address the public health threat posed by the tobacco epidemic and, in parallel, create the conditions for the development of sustainable alternatives to tobacco farming suitable for smallholder agriculture. The policy implications and related recommendations are fourfold.

- **Strong domestic tobacco-control policies.** These can be pursued aggressively by all tobacco-exporting and importing nations, knowing that global demand for tobacco leaf is likely to continue at current or higher levels for some time. This demand shelters the livelihoods of the current generation of tobacco farmers from domestic policy change.

- **Strategic public investment in market infrastructure, technical assistance and access to financing.** Evidence from a variety of settings shows that many other crops, crop combinations, farming systems and livelihood strategies can offer better opportunities for farmers. However, the market guarantee provided through contract farming and supply management remains compelling for farmers with limited access to market infrastructure and financing. The challenge for a tobacco-farming transition is to overcome the efforts of many years of decline in agricultural markets and limited government investment in smallholder agriculture. Access to public resources is necessary, especially in the form of credit financing, technical assistance and the development of local and regional market infrastructure. As adjustment to new cropping systems and product markets takes time, creativity and resources, government support for access to start-up capital and the incremental costs of the transition period is also justified. While this is not easy for resource-poor governments, much can be done through different ministries and levels of government by coordinating the use of existing agricultural policies and programs. National food acquisition and school feeding programs designed to provide a social safety net for the food insecure while also stimulating smallholder food production are examples. It can also be financed in part by the gradual withdrawal of public funding for tobacco farming.
- **Diversification, not substitution.** While substitute crops are likely to be essential components of economically sustainable alternatives to tobacco growing, single crops are not magic bullets and are unlikely to be the answer for smallholders in most regions where tobacco is currently grown. The reasons for this are twofold. First, it is technically difficult and costly to simply insert another monocrop into the same space as tobacco and in the same season without first improving the soil. Tobacco is notoriously hard on the soil, not only depleting nutrients and soil organic matter but also exacerbating weed infestation, soil erosion, soil pollution and salinization. Second, a focus on the direct substitution of another crop limits the range and diversity of cropping systems and livelihoods options farmers can consider. A focus on crop and livelihood diversification rather than a single substitute crop offers more scope for innovation and the use of all resources available, including minor seasons, marginal soils, value-added production and off-farm employment.
- **A national vision for sustainable agriculture.** Efforts to foster a transition out of tobacco farming cannot be disconnected from the national system of food production, agricultural land use and rural development within which tobacco farming is embedded. The experience of transition in high-income countries suggests that alternatives to tobacco farming framed

only in terms of the international competitiveness of single agricultural commodities would likely displace smallholder farmers from the land, not help them develop as food and income producers. Furthermore, the single criterion of international competitiveness reduces agricultural policy to trade policy rather than support for the more complex function agriculture plays in most low- and middle-income countries such as food security, the sustainable use of natural resources and the development of local enterprise. A national vision for sustainable agriculture, and efforts to overcome the effects on smallholder agriculture of decades of declining government support are consequently key to the tobacco-farming transition.

Developing a national vision for sustainable agriculture, promoting livelihood diversification and investing in market infrastructure, technical assistance and access to financing are complex policy undertakings. Article 17 of the FCTC provides a constructive and modest space for the tobacco-control community and health ministries to play a supportive role to ministries of agriculture, the environment, labor, finance and rural development with the mandate to pursue these goals. It can also help the tobacco-control community build the alliances needed to counterbalance the current influence of the tobacco industry on government policy, and create buy-in from tobacco farmers engaged in the legitimate search for better livelihoods. In this process, tobacco farmers should not be blamed for growing tobacco or treated as passive victims of an all-powerful industry. Rather, the task is to engage the tobacco-farming community and other stakeholders in the development of better options. Only then can governments set themselves on a track towards a fair and orderly transition consistent with the goals of tobacco control and the spirit of the FCTC.

CONTRIBUTORS

Farida Akhter is the executive director of UBINIG, a policy research organization in Bangladesh. She has been involved in community-based action research for 30 years and is an organizer with the national farmers' movement Nayakrishi Andolon. A leading advocate of women's rights in Bangladesh, Ms. Akhter has published in English and Bangla on a wide range of issues affecting women's health, the environment and women in agriculture.

Daniel Buckles is an independent consultant and adjunct research professor at Carleton University (Ottawa, Canada). He has published various books on the social and economic dimensions of agriculture, including *Cover Crops in Hillside Agriculture: Farmer Innovation with Mucuna.* (CIMMYT/IDRC) and *Food Sovereignty and Uncultivated Biodiversity in South Asia* (Academic Foundation/IDRC). Recent publications focus on methods for engaging communities in action research, including *Fighting Eviction: Tribal Land Rights and Research-in-Action* (2013, Cambridge University Press India) and *Participatory Action Research: Theory and Methods for Engaged Inquiry* (2013, Routledge).

Jad Chaaban is an associate professor of economics at the American University of Beirut in Lebanon. His primary research interests are agriculture and development economics. He has published numerous book chapters and journal articles on food security, agricultural rural development and the economics of tobacco control.

Guilherme Eidt Gonçalves de Almeida is a lawyer and independent consultant with experience in international public law, human rights and agrarian, environmental and health law. He currently studies public policies affecting health and the environment, rural sustainable development and territorial development. He has been active in the tobacco-control movement in Brazil for many years and is author of the book *Fumo: Servidão Moderna e Violações de Direitos Humanos* [Portuguese, Tobacco: modern servitude and human rights violations] published in 2005 by Terra de Direitos.

Laura Graen is an anthropologist and campaigner with the consulting firm Research for Changemakers (www.forchangemakers.com) and a co-founder of Unfairtobacco.org. Her work focuses on research and campaigning for non-governmental and inter-governmental organizations dedicated to solving pressing socioeconomic challenges in the developing world.

Kanj Hamade holds a PhD in agricultural economics from the University of Bologna (Italy) and is currently an Adjunct Assistant Professor at the Lebanese University (Beirut, Lebanon). His research focuses on rural development practices, institutional economics and mixed methods research.

Rafiqul Haque Tito is a regional coordinator for UBINIG, a policy research organization in Bangladesh. Based in Cox's Bazaar, a coastal region of Bangladesh, he has mobilized thousands of farmers in support of biodiversity-based farming and the conservation of mangrove forests, and is an accomplished organizer with the national farmers' movement Nayakrishi Andolon.

Jacob K. Kibwage is an associate professor and founding dean of the School of Environment and Natural Resources Management at the South Eastern Kenya University (Kitui, Kenya). His research interests include environmental studies and sustainable livelihoods, including the promotion of economically viable alternatives to tobacco farming. He has contributed directly to various international technical forums on tobacco issues, included the WHO policy guidelines on Articles 17 and 18 of the Framework Convention on Tobacco Control.

Natacha Lecours is a program management officer for the Non-Communicable Disease Prevention program at the International Development Research Centre (IDRC). A social scientist with expertise in environmental and health sciences, her research focuses on challenges faced by farming communities across low- and middle-income countries in South America, Africa and Asia.

Wardie Leppan is a senior program specialist in the Non-Communicable Disease Prevention program of the International Development Research Centre (IDRC). He has been active supporting research in the field of tobacco control at IDRC since 1998, and fulltime since 2004. One of his key areas of focus has been on tobacco farming. He participated in a number of meetings of the FCTC Ad Hoc Working Group on Alternative Crops along with a number of IDRC partners.

Peter O. Magati is a lecturer at the School of Finance and Applied Economics at Strathmore University in Nairobi, Kenya. His current doctoral research focuses on the economics of tobacco production, building on prior studies in economic and social research, strategy planning and policy analysis.

Godfrey W. Netondo holds a PhD in botany and is a professor in the Department of Botany, School of Biological and Physical Sciences, Maseno University, Kenya. His research specializes in plant stress eco-physiology, within the broader field of agrobiodiversity conservation and ecosystem services.

Marty Otañez is an assistant professor in the Anthropology Department, University of Colorado at Denver. He uses methods from visual ethnography to support research on labor rights, corporate accountability and global public health policy making. He is a board member of the Human Rights and Tobacco Control Network (www.hrtcn.net) and operates the blog www.fairtradetobacco.org.

Lightning Source UK Ltd.
Milton Keynes UK
UKOW04n0016261014

240614UK00004B/94/P